MW00911092

Into Thin Air

SCIENCE AND SOCIETY

Into Thin Air

THE PROBLEM OF AIR POLLUTION

J. S. Kidd and Renee A. Kidd

Facts On File, Inc.

This one is for Loren.

Into Thin Air: The Problem of Air Pollution

Facts On File, Inc.
11 Penn Plaza
New York NY 10001

Library of Congress Cataloging-in-Publication Data

Kidd, J. S. (Jerry S.)
Into thin air: the problem of air pollution / J.S. Kidd and Renee A. Kidd.
p. cm.—(Science and society)
Includes bibliographical references and index.
Summary: Examines the causes of atmospheric pollution, acid rain, ozone depletion, and global warming and explains how these conditions affect our health and economic prosperity.
ISBN 0-8160-3585-7
1. Air—Pollution—Juvenile literature. 2. Climatology—Juvenile literature. [1. Air—Pollution. 2. Pollution. 3. Environmental protection.] I. Kidd, Renee A. II. Title. III. Series: Science and society (Facts on File, Inc.)
TD883.13.K54 1998
363.739'2—dc21 98-9886

Facts On File books are available at special discounts when purchased in bulk quantities for businesses, associations, institutions or sales promotions. Please call our Special Sales Department in New York at 212/967-8800 or 800/322-8755.

You can find Facts On File on the World Wide Web at http://www.factsonfile.com.

Text and cover design by Cathy Rincon

Illustrations on pages 2, 6, 15, 18, 21, 23, 36, 47, 50, 65, 66, 89, 100, and 109 by Jeremy Eagle

Printed in the United States of America.

MP FOF 10 9 8 7 6 5 4 3 2 1

This book is printed on acid-free paper.

Contents

Preface

The present book is one in a series on the general theme of science and society. *Into Thin Air* is about the attempt to solve a major societal problem—the control of air pollution.

Controlling air pollution is not limited to suppressing harmful emissions. Although atmospheric chemistry is the key area of research regarding air pollution, an understanding of weather science is also vital. Air dilutes emissions, so the physics of air movement and the technologies of weather forecasting and weather modification are also key sources of knowledge.

Indeed, weather science makes its own major contribution to the well-being of the world's population. New technologies provide the five- and seven-day weather forecasts now available through radio, television, newspapers, and the World Wide Web. High-speed computers, orbiting satellites, and precision radar make it possible to track weather events. Weather scientists also give warnings of catastrophic weather conditions such as hurricanes and tornadoes. Many lives are saved when the affected areas are evacuated and emergency personnel are alerted.

The field of climatology is also important in combatting pollution. Climate studies reveal seasonal and long-range weather trends such as an area's projected yearly rainfall. From this information, atmospheric chemists and environmental scientists can calculate the extent to which pollutants will be washed out of the air and into lakes and streams.

Scientists now know that some forms of air pollution have the potential power to change the climate. This possibility expands public interest from concerns about local pollution to concerns about the worldwide contamination of the atmosphere.

In accord with the basic theme of this series, this volume explores the contributions of science to societal goals such as good health and economic prosperity, as well as a basic understanding of the causes and effects of atmospheric pollution. It also covers the ways in which scientific information can support the decisions of public officials in response to the problem. In addition, this book explores the means by which science is influenced and supported by governmental institutions and society in general.

The key federal government agencies concerned with the problem of atmospheric pollution include the National Oceanic and Atmospheric Administration (NOAA), the U.S. Weather Service (a part of NOAA), the Department of Energy, the National Aeronautics and Space Administration (NASA), and the Environmental Protection Agency (EPA). Other influential institutions include units of state and local governments, commercial and industrial firms, and academic institutions. Multinational bodies such as the European Union (formerly the Common Market) have units that are responsible for collective action to control pollution. On the global level, the United Nations organization provides channels for even wider participation among nations. The services provided to member nations include collection and distribution of technical climate information and meetings that promote international agreements on global pollution problems.

As characteristic of other volumes in the series, this book discusses the relevant areas of scientific study and the pertinent discoveries that arose from such studies. The book also introduces some of the people who developed the relevant ideas and techniques. It shows how technical and political problems were met and sometimes overcome. This volume also attempts to convey a sense of the struggle and adventure that is part of all scientific endeavor.

Acknowledgments

We thank the faculty and staff of the College of Library and Information Services at the University of Maryland, College Park—particularly Ann Prentice and Diane Barlow. Professor Alistair B. Fraser at the Pennsylvania State University gave us some cogent advice as did Dick Feely at the Pacific Marine Environmental Laboratory, National Oceanic and Atmospheric Administration, Seattle, Washington. Gwen Pitman of CCI, Inc., and Amy Ballard of the Smithsonian Institution helped us find useful images. We also want to offer special thanks to the librarians at the University of Maryland and at the National Research Council in Washington, D.C., for their consistent support.

1

The Atmosphere

Atmospheres are composed of gases. These gases are sometimes made up of single atoms such as neon but more often are composed of compound molecules such as natural oxygen, which consists of two oxygen atoms. In addition to oxygen, the air we breathe contains nitrogen molecules, made of two atoms of nitrogen. (Its symbol is N_2.) (The small number after the elements' symbols indicates how many atoms of that element are necessary to form the molecule.) Some of the gases in the earth's atmosphere are made up of two different elements. Carbon dioxide (CO_2), for example, is composed of one atom of carbon joined with two atoms of oxygen.

Atmospheres in the Solar System

The moon, our nearest astronomical neighbor, cannot hold an atmosphere because its small size does not provide sufficient gravity to hold even heavy gases. The gravitational pull of our small moon is no match for the mighty tug of the vacuum of interplanetary space. Any gases that might have been present

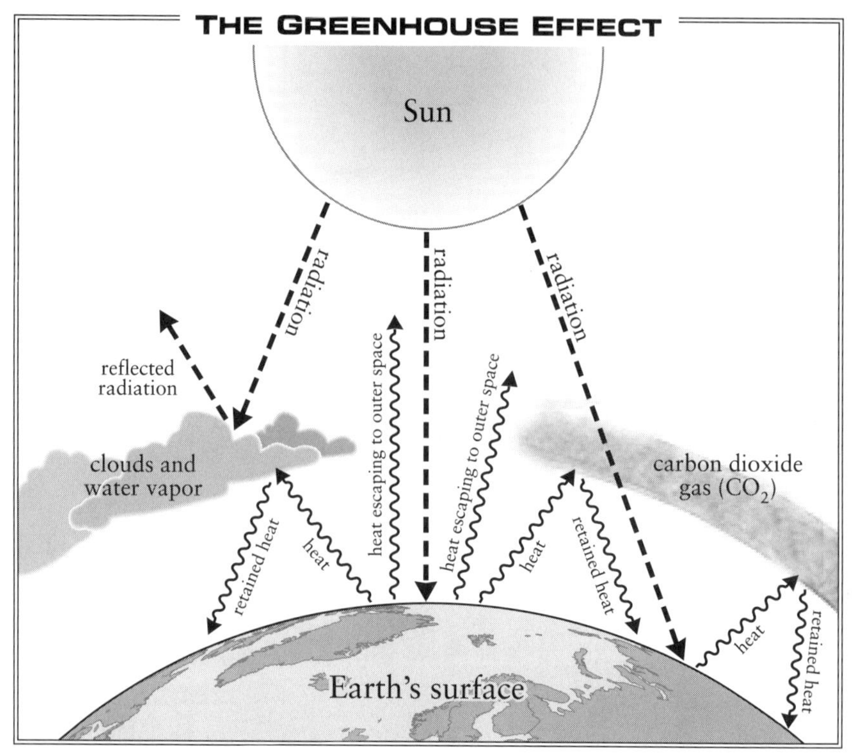

The greenhouse effect is caused when CO_2, water vapor, and other gases block the escape of radiant heat from near the earth's surface.

when the moon was formed were sucked into deep space millions of years ago.

Mars, called the Red Planet, has a very thin atmosphere. The planet is barely large enough and dense enough to hold a wispy mixture of gases. However, Venus, known as the Morning Star, is large and dense enough to hold a thick layer of gas on its surface. This gas is a toxic mixture of carbon dioxide with a dash of nitrogen and some sulfur compounds. At surface level, the gas is hot enough to melt lead. This superheated temperature exists because Venus is nearer to the Sun than Earth is. In addition, the thick layer of carbon dioxide gas traps much of the sun's radiant energy and keeps the intense heat close to the surface of Venus.

This process is known as the "greenhouse effect." Most of the sun's rays pass through the carbon dioxide layer and reach the surface of the planet. The surface absorbs some of the energy and generates heat. The heat is reradiated skyward, but the carbon dioxide blocks the upward movement. The CO_2 holds the heat and sends it back toward the surface. Therefore, the heat energy is trapped between the CO_2 layer in the atmosphere and the surface of the planet.

In contrast to its neighbors, Earth's atmosphere is about 80% nitrogen. The nitrogen molecule, N_2, is a very placid molecule that does not react easily with other chemicals. Oxygen and small traces of other gases make up the other 20% of the atmosphere. These trace gases include a small amount of carbon dioxide. The carbon dioxide holds enough heat energy near the surface of the earth to allow the planet to support life. This trace amount of CO_2, about .003%, is far less than that in the Venusian atmosphere.

Earth's Atmospheric Changes

Earth's atmosphere has evolved over time and continues to do so. The original atmosphere of the hot, newborn Earth was probably a mixture of hydrogen and helium. The hydrogen and helium atoms are the smallest of all the elements. Molecules composed of these elements are lighter than air—think of how a helium-filled balloon will quickly rise skyward. As the earth cooled, hydrogen and helium were sucked away by the vacuum of outer space.

During the next stage of atmospheric evolution, heavier gases—ammonia, containing nitrogen and hydrogen, and methane, containing carbon and hydrogen—were the main components. These heavier gases arose from deep within the earth and escaped from volcanoes, vents, and geysers. Scientists believe that such gases provided the raw materials for the first life on Earth. A current theory suggests that life originated in water containing molecules of ammonia and methane. The development of life was activated by sunlight and, perhaps, lightning. According to this theory, the first life-forms were probably simple molecules assembled from the ammonia and methane. These molecules were able to reproduce themselves. Eventually, such molecules formed a protective skin or membrane and became one-celled creatures.

There was little, if any, oxygen in that second period of the earth's atmosphere. In fact, atmospheric oxygen probably would have been poisonous to these primitive creatures. Over eons of time, the atmosphere changed and the proportions of oxygen and molecular nitrogen increased as the reaction of ammonia and water released both gases. The ammonia (NH_2) yielded the nitrogen that dominates the present atmosphere. The water (H_2O) gave up part of its oxygen in the process. The methane (CH_4) was transformed into carbon dioxide (by bonding with oxygen). Other carbon-based molecules were washed out of the air by rain. Slowly, the atmosphere became richer in oxygen.

In the meantime, some of the primitive life-forms that existed in the ammonia-and-methane period migrated to an oxygen-free environment in the deep sea or far underground. Their successors still exist in such places. Other single-celled creatures adapted to the gradually increasing proportion of oxygen in the air. Those that adapted to oxygen became the forerunners of all the multicelled plants and animals that now occupy the earth.

The Seasons

Scientists are not always serious. One atmospheric chemist remarked that the earth enjoys the "Goldilocks" condition. Our planet is not too hot, not too cold, but just right. Of course, that scientist was referring to ideal conditions for life. An essential life-supporting condition is the availability of liquid water. Water (H_2O) comes in three forms. Solid water is ice, gaseous water is steam or water vapor, and liquid water is the substance that is needed for life. Ice becomes liquid when it is heated, and liquid water becomes a gas when it evaporates. The evaporation of a liquid is hastened by heat, and steam forms when the temperature of the liquid is raised to its boiling point (240 degrees F, 100 degrees C). On some planets, the cold is so intense that water (if there is any) is frozen solid. On other planets, the temperature is so hot that all water (if there is any) is in the form of water vapor or steam.

In our solar system, Earth appears to be the only planet that can sustain liquid water. There may be traces of water on Mars, but this water would be frozen solid. Mars has too little atmosphere to hold the sun's heat and, therefore, is much too cold for liquid water. Some of the large moons that circle Jupiter and Saturn may have liquid water below a frozen surface. The telescope on the *Galileo* spacecraft has detected what appear to be ice floes on Europa, one of Jupiter's moons. If these ice floes indicate the presence of liquid water under a covering of ice, the warmth to keep it liquid is not supplied by the distant sun. The heat probably results from long-term radioactivity. This action is similar to the process that forms lava by melting the rock below the earth's solid crust.

In contrast to the other planets in the solar system, the average temperature of Earth is in the moderate range. This condition allows liquid water to exist over most of its surface. Although Earth's average temperature may be "just right," many places on Earth experience temperatures that vary greatly. Local weather conditions can change from hour to hour and as well as during seasonal weather cycles.

The seasons occur because the earth is tilted on its axis. As the earth circles the sun, the direction of the tilt always remains the same. Winter comes to the Northern Hemisphere when the tilt of the axis moves the northern part of the earth away from the sun. Conversely, summer arrives when the northern half is tilted toward the sun. The seasons in the Southern Hemisphere are always the reverse of those in the Northern Hemisphere. When it is winter in New York, it is summer in Rio de Janeiro and vice versa.

In past times, those who studied geography divided the earth into various regions. The surface of the earth was divided into the Southern and Northern Hemispheres by an imaginary line

The Revolution of the Earth and the Seasons

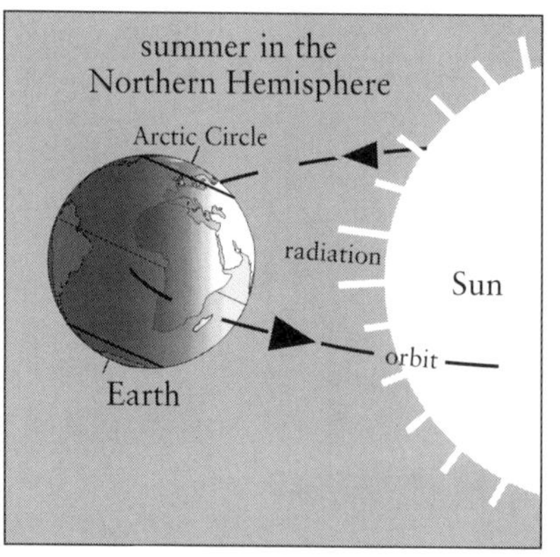

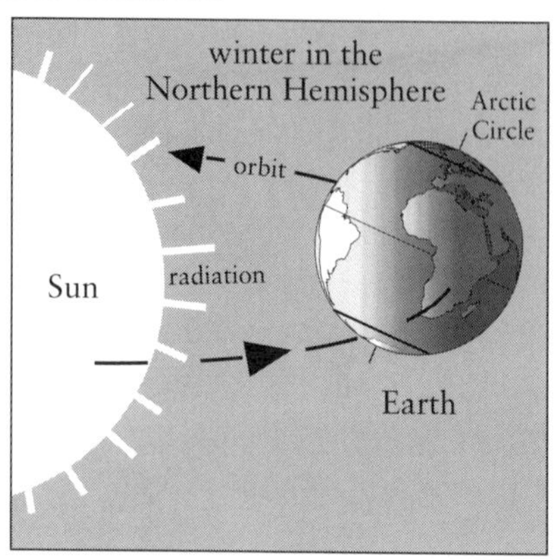

The Earth is tilted relative to the plane of its orbit around the Sun. Consequently, during the winter in the Northern Hemisphere the Arctic is dark in the daytime and the Antarctic region is lighted by the sun even at night. When it is summer in the Northern Hemisphere, the opposite conditions exist.

called the equator. The equator is midway between the North Pole in the Arctic and the South Pole in the Antarctic. Three climate zones lie on each side of the equator. Tropical climate zones are found nearest the equator. Except near the tops of tall mountains, temperatures in the tropics rarely, if ever, fall below the freezing point of water. The temperate zones range between the tropics and the coldest areas. The temperate zones have the most pronounced seasonal changes in temperature, rainfall, and other variables. However, the average temperature is mild—about 60 degrees Fahrenheit or 15 degrees Celsius.

The zones farthest from the equator and surrounding the North or South Poles are the Arctic zones. *Arctic* is a Greek word referring to the north. Because the scholars who wrote the original geography books lived in the Northern Hemisphere, they called the southern polar region the "Antarctic" or opposite to the Arctic. In the Arctic zones the temperature is consistently below the freezing point of water and most snow in the Arctic zones remains frozen. Each year, the previous snowfall is compacted into ice by the weight of the new snow. Thus, the Arctic zones are said to have a permanent ice cap.

In the tropical regions nearest the equator, there is no appreciable winter season. However, other changes take place in a rhythmic fashion. In the Caribbean area, the hurricane season starts in late June and lasts until November. Central America has a dry season from November to April and a wet season from May to October with a short dry spell in July.

Other areas of the globe experience yearly cycles of torrential rains known as monsoons. In India, for example, the monsoon season lasts from June through September. The rainy months are followed by a warm autumn, a mild winter, and a hot period from March through May.

Climate and Climate Change

All these rhythmic variations take place within a set of conditions called climate. The idea of climate represents a combination of factors such as temperature, humidity, and rainfall that follow general patterns in a given region. In what is described as a Mediterranean climate, the days are quite warm, the nights are chilly, and the temperature rarely falls to the freezing point of water. San Francisco, California, has a Mediterranean climate with the added peculiarity that the summer months may be slightly colder than the winter months because of shifting ocean currents.

In the study of climate, average temperature and average rainfall are the most important factors. Ireland provides a good example of a truly temperate climate. The range of temperatures is narrow. In the summer, the temperature rarely rises above 80 degrees Fahrenheit (27 degrees Celsius). When it does, the Irish are convinced that they are suffering from a heat wave. In the winter, some ice may form on still water at night, but water rarely freezes during the daytime. Indeed, year-round temperatures usually range between 50 and 70 degrees Fahrenheit

(10–21 degrees Celsius). Although the temperature range is low, the amount of rainfall is high. The areas covered by natural vegetation remain green all year. Consequently, Ireland is called the Emerald Isle.

The average temperatures found in midwestern areas of the United States are somewhat misleading. Fairly extreme fluctuations frequently occur. For example, the temperature in central Indiana can reach more than 100 degrees Fahrenheit (38 degrees Celsius) in the summer and 10 degrees below zero Fahrenheit (minus 23 degrees Celsius) in the winter. Yet, Indiana is considered to have a temperate climate with an average temperature of about 51 degrees Fahrenheit (11 degrees Celsius). Thus, although the average temperature is temperate—a mild, middle value—the extremes can be either tropical or arctic. In fact, the temperature in Indiana can change by 40 degrees Fahrenheit (4 degrees Celsius) overnight.

Scientists now fear that our changeable climate will change even more dramatically. The world faces the prospect that human activities are responsible for unpleasant weather modifications. Climatologists fear that the blanket of carbon dioxide gas found in the Venusian atmosphere may someday have a counterpart on Earth. As yet, Earth's blanket of carbon dioxide is much thinner than the one that keeps Venus unbearably hot. However, air pollution is slowly thickening the blanket that surrounds Earth.

This condition might cause global temperatures to rise by several degrees over the next 50 years. Indeed, the warming trend might be sufficient to melt a large portion of the polar ice caps and cause sea levels to rise. The trend could cause tropical conditions to move northward in the Northern Hemisphere and southward in the Southern Hemisphere. The temperate zones would become warmer. Agricultural conditions would change around the world.

Such temperature variations due to carbon dioxide build up can occur as a consequence of burning oil and coal, the fossil fuels. Fossil fuels are the residue of plants and animals that died millions of years ago. Coal is mainly a byproduct of long-dead plant matter. When the plants died, their remains were gradually buried deeper and deeper by layers of earth. Plants are composed mainly of carbon compounds, and the pressure of the overlying earth slowly compressed the carbon into coal. Although some coal contains both carbon and impurities, hard coal is almost pure carbon. Carbon is inflammable and burns with a very hot flame. Burning coal releases CO_2 into the atmosphere.

The principal fuel for the production of electricity is oil. Oil, too, is a fossil fuel. It comes from the remains of dead microscopic creatures that were buried under layers of earth. Oil is not a solid like coal, but a liquid form of carbon. Oil is the raw material from which gasoline is made. Burning oil and gasoline also releases CO_2.

Some atmospheric carbon dioxide is absorbed by green plants. However, most is dissolved into the water of the oceans. A small part of that carbon dioxide is used by sea creatures. They combine it with calcium to manufacture calcium carbonate. The sea creatures use the calcium carbonate to form their shells. Later, these shells fall to the sea floor. Thus, the activity of some of these small animals captures a minor portion of atmospheric carbon dioxide and retains it for very long periods of time.

In recent years, the production of carbon dioxide has outstripped the absorption capabilities of the seas, the sea animals, and the green plants. The portion of carbon dioxide that is not absorbed is retained in the atmosphere.

Climate Cycles

The climate of the earth has changed many times in its 4.5-billion-year lifespan. The positions of the continents have also changed during that time. In the past 400 million years, the continents have come together in a single mass and then separated at least twice. As these land masses have shifted over the surface of the earth, they have passed through different climate zones. Scientists have found evidence that the frozen wastes of Antarctica were once a green and pleasant place.

Louis Agassiz, a Swiss-American naturalist, was one of the first scientists to study long-term changes in climate. In 1836, he noted the positions and markings on many large rocks found in the lower Alpine valleys of Switzerland. A few decades earlier, two Swiss scientists had noted that these boulders were identical to rocks embedded in high Alpine glaciers. Agassiz deduced that earlier glaciers had carried the rocks to the lower parts of the valleys. He also noted that the bedrock in valleys and on the lower slopes of mountains was marked with many parallel grooves. From these and other observations, Agassiz reasoned that in times past, the Swiss glaciers had been much larger.

Agassiz observed similar features in stretches of rock found in northern Europe and England. He came to believe that an ice sheet had once covered a large portion of the European continent. Most scientists were at first skeptical of his ideas. However, scholars from all over Europe were soon convinced by Agassiz's careful investigations and his excellent reputation. The following year, he expanded this theory. He proposed the existence of an ice age during which ice sheets had extended over northern Europe, Asia, and North America. This theory was too fantastic for his fellow scientists, and they laughed at his ideas. Nevertheless, Agassiz continued his research. He pointed out that the grooves in the areas of flat rock were parallel, continued for many yards, and were all oriented in a north-south direction. Agassiz believed that the deep cuts were made by stones carried along by the glaciers as they flowed over the bedrock. He also pointed out that mounds of dirt and stones were piled up where the grooves stopped. Agassiz reasoned that the glaciers had acted like bulldozers and pushed along debris to the farthest reaches of the ice sheets. Eventually, enough evidence accumulated to prove Agassiz's theory of extensive glaciation. By 1845, most scientists had accepted the concept of an ice age.

Louis Agassiz went on to become one of the most renowned naturalists in the history of science. In 1846, he visited the United States. The following year, Agassiz joined the science faculty at

The Barnard glacier in Alaska has multiple sources and visibly carries rocks and dirt from the mountain sides as it flows toward the sea. (*Austin Post* photograph, July 29, 1957, courtesy of the National Snow and Ice Data Center)

Harvard College (now University) in Cambridge, Massachusetts. He explored much of North and South America and recorded his observations of plants, fish, and animals. Agassiz was the moving force behind the establishment of the National Academy of Sciences in 1863.

New Forces in Climate Change

The transition period between cold and warm climate cycles generally covers several hundred years. However, humans are now confronted with the possibility that a serious climate change might take place in only a few decades. If so, major adjustments in agriculture, fishing, and forestry would be required.

The prospect of rapid climatic changes is the direct result of human activities. In the recent past, many people were totally unaware that their actions could prove destructive to the environment. People burned many things—coal, trash, leaves, and garbage—to heat their homes, cook their food, and keep their cities and farms free from debris. Not many years ago, children enjoyed the autumn treat of roasting marshmallows over a bonfire made from fallen leaves. No one thought that these activities would damage the atmosphere. Now, almost everyone knows that many of these activities are harmful. However, not everyone has been willing to change their behavior. People persist in following their past habits. For example, to save time and money, some factory owners continue to allow noxious gases to pollute the air. Because of these behaviors—whether innocent, thoughtless, or intentional—humans may be forced to make adjustments in the way they obtain food, clothing, and shelter in the near future.

2

The Weather

In England, people say, "If you don't like the weather, just wait a few minutes and it will change." Daily or even hourly change in the weather is a feature of the temperate zones. Some of these weather changes cause moderate discomfort. Cold rain is chilling, and wind makes it chillier. However, these conditions are not disastrous, and people can usually cope with them. For example, most people in England carry an umbrella if they plan to be outside for any length of time. On the other hand, some weather—snow, hail, or fog—produces dangerous conditions. Driving a car or just walking is hazardous. Growing crops can be flattened by hail. Major storms, hurricanes, and tornadoes destroy property and can cause fatalities for humans and animals.

Early Weather Prediction

At the dawn of history, weather was associated with religion. Ancient people looked to the gods for an explanation of the lightning and thunder generated by violent storms. Zeus, the "father" of the Greek gods, was originally a sky and weather god. He is often shown holding a bolt of lightning.

Weather prediction was based in part on superstitions. Human beings saw the flight of birds or the agitation of barnyard animals as forecasts of the gods' weather plans. Many people thought they could predict an unusually cold winter by the thickness of an onion skin or the density of a sheep's wool.

Although daily weather patterns were difficult to discern, seasonal changes did show a predictable pattern. This pattern was correlated by ancient observers with the heavenly locations of constellations and planets. Weather science began when these patterns were recorded and used for systematic predictions.

The ancients regarded the celestial changes as calendars and fixed their annual ceremonies by the heavenly configurations. Stonehenge, a circular monument of huge stones on Salisbury Plain in England, was associated with one of the most famous seasonal rituals, the spring equinox celebration. The placement of the stones is correlated to the position of the sun at the onset of spring.

Egyptian astronomers were the first to establish a one-year calendar of 365 days. Their calendar was probably based on observations of two very special events. These events happened each year within a few hours of each other. At dawn on the day that the Nile usually overflowed its banks, the sun ascended into the sky, accompanied by the Dog Star, Sirius. The yearly flooding of the Nile brought precious water and new soil to the lands along the river and was vital to Egypt's agriculture. Therefore, the yearly cycle of life began with this event.

The early Greeks saw other weather patterns. They remarked on the connection between wind direction and temperature. About 350 B.C., the Greek philosopher Aristotle sought to study weather in a systematic manner. He observed that warm, moisture-laden air rose upward and formed clouds when it reached a high altitude. Aristotle's study of clouds, stars, planets, and other heavenly bodies led to his book *Meteorologica*, the study of "things above." Meteorology, the modern name of weather science, is based on this Greek word.

In the hundreds of years between Aristotle's time and the beginning of the 20th century, weather study and weather prediction remained at the level of folk knowledge. Some of this lore was helpful for the farmers and sailors who depended on the weather. Keen observers could read cloud patterns and make fairly accurate, short-range predictions. Halos around the sun or the moon were reasonable indicators of coming storms. "Red sky in morning, sailor take warning—red sky at night, sailor's delight" is not a bad guide to a day's upcoming weather.

In the 1500s and 1600s, European scholars began to make significant improvements in weather science. Around 1500, during the High Renaissance, Leonardo da Vinci, an Italian artist and scientist, constructed a superior weather vane. Galileo, a famous Italian astronomer, invented the thermometer in 1593. His student Evangelista Torricelli made an instrument for measuring air pressure in 1643. This device later evolved into the modern barometer.

In 1653, Ferdinand II, grand duke of Tuscany, established several weather recording stations in the area north of Rome.

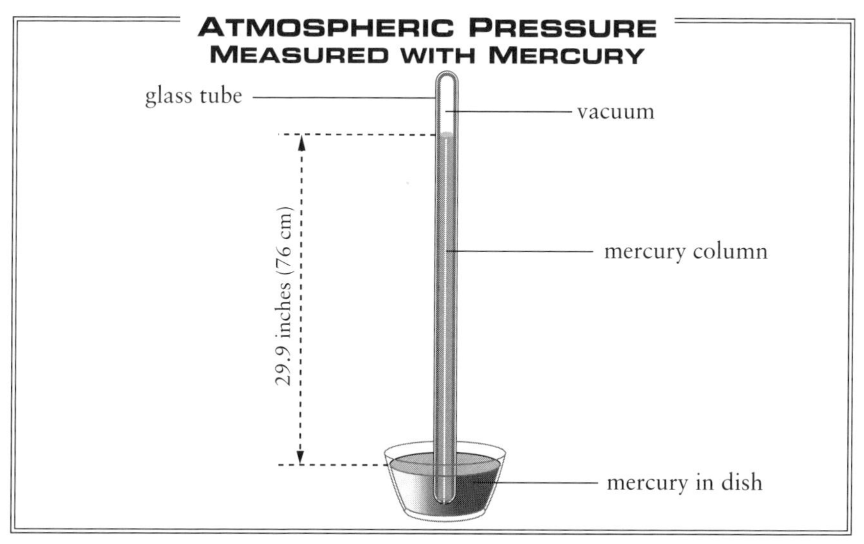

Atmospheric pressure pushes the mercury column 29.9 inches (76 centimeters) above the surface of the mercury in the dish. The column of mercury will climb slightly higher or fall slightly lower when the weather changes.

This project was the first to prove that similar weather patterns occurred in different locations at the same time. The duke also investigated the possibility that weather conditions in one location could be used to predict conditions at other locations.

In the early 1660s, Robert Boyle, an English chemist, began to study the physical nature of gases such as air. He was the first to show the relationship between pressure and the volume of a gas. He showed that air could be compressed and that it could also be removed from a container to make a near vacuum.

Around 1800, the French physicist Jacques Charles established the relationship between the compression of a gas and its temperature. His investigations began in 1783 while using hydrogen balloons to test the possibilities of human flight. His theory states that a gas becomes warm when compressed under pressure and cools after the pressure is released and the gas is allowed to expand.

The work of these scientists resulted in a series of studies that led to the belief that weather was the result of physical laws. If the belief were true, accurate weather predictions would be possible. Scientists hoped that meteorology would become an exact science by applying the proper physical laws to accurate measurements of temperature, wind speed, rainfall, and other variables. They were disappointed by the actual turn of events.

Forecasting

The British, French, and Spanish colonists who occupied North America in the 1600s were shocked by the violence of the winters. The eastern seaboard of North America—where most of these western Europeans settled—has a "continental climate." That is, the weather conditions experienced on the east coast develop over dry land to the west and are borne eastward by the prevailing winds. In contrast, European weather develops over the Atlantic Ocean and is affected by the mass of warm water carried by the current called the Gulf Stream to the northeastern Atlantic Ocean.

The Gulf Stream gets its warmth and its name from the Gulf of Mexico, where it originates. It carries warm water from the tropics along a path that is parallel to the east coast of the United States. Then it cuts across the Atlantic to the west coast of Europe where it disperses. Because of the presence of warm water off the west coast of Europe, those coastal regions have a warmer winter than the northeastern United States. Parts of eastern Europe, such as Poland and Russia—far from the warm Gulf Stream—experience the same harsh winter weather conditions as eastern North America.

By the late 1700s, observing weather conditions was a favorite hobby of prosperous Americans. George Washington, Thomas Jefferson, and Benjamin Franklin were avid weather watchers. They routinely recorded their observations of temperature, precipitation, wind conditions, and barometric pressure. George Washington made such notations on the day before he died in 1799.

Benjamin Franklin was involved in more active investigations of weather conditions. During an electric storm in 1751, he tied a key to a kite string. The position of the string allowed the key to touch the ground. He then flew the kite into the high wind of a thunderstorm. When a bolt of lightning struck the kite, the lightning traveled down the wet string. The string acted like a wire to conduct the electricity through the key into the ground. Franklin had shown that lightning and electricity were the same phenomenon. He survived his foolhardy investigation, but other men were electrocuted when they tried to duplicate his experiment.

In those colonial days, the scientific aspects of weather investigation were incorporated into a broad area of learning called natural history. Later, during the mid-1800s when scientific studies became more specialized, the field of meteorology was classified as a physical science. Much of the research, however, was in the hands of talented amateurs.

William Redfield, a prosperous New Yorker, studied meteorology as a hobby. Redfield examined weather charts that had been compiled over several years. He discovered that weather systems have both a circular movement and a curving, sideways movement. The circular movement accounts for wind direction, while the sideways movement indicates the direction in which the weather system is moving. If one knew the characteristics of the particular weather system, one could predict the effects along its path. Unfortunately, there was almost no communication between amateur meteorologists such as Redfield and the astronomers and technicians who tried to forecast the weather. Redfield's important discoveries received little attention at the time.

In the early 1800s, the U.S. government did not sponsor weather research. Politicians had little interest in scientists who theorized about the weather. They understood, however, the economic importance of being able to warn farmers and fishers of impending storms. Nevertheless, politicians were hesitant about spending money on weather research.

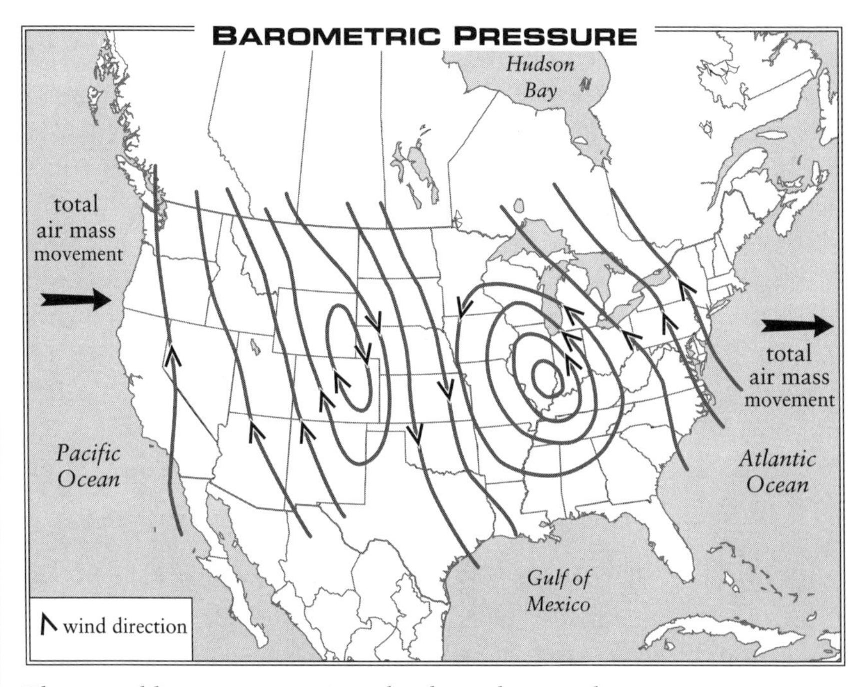

The curved lines connect points that have the same barometric pressure. The succession of lows and highs moves gradually from west to east across the continent.

Although forecasters had few solid scientific theories or systematic observations at that time, they recognized that weather conditions in the United States usually originate in the west and travel toward the east. Even if weather information had been more scientific, weather forecasting would have improved very little. At the time, there were no fast communication systems to carry the weather observations.

The problem was partly resolved in 1832 when Samuel F. B. Morse, an American artist and inventor, filed a patent for the telegraph. Wires soon linked railroad stations and post offices from the Midwest across the Allegheny Mountains and on to Washington, D.C.

Joseph Henry, the first secretary (chief executive) of the Smithsonian Institution, used this technological advance to support a systematic weather service. By 1860, he had recruited 500 volunteers from various locations between Washington, D.C., and the Mississippi River. Every day, these volunteers used the telegraph to report weather observations to an office at the Smithsonian. The information was summarized, forecasts were composed, and the predictions were reported in the evening newspapers. Unfortunately, people whose work was most affected by the weather—farmers and sailors—did not have access to the newspapers. Communities near Washington, D.C., tried several methods to speed the flow of information. Their systems, such as raising signal flags on the roofs of post offices, were not very successful. Although Henry's concept was a step in the right direction, the predictions were not very accurate. The outbreak of the Civil War disrupted his system, and it was shut down in 1863.

Military operations during the Civil War demonstrated that weather information could be a valuable asset to commanders in the field. Consequently, after the war ended, former military men elected to the House of Representatives worked to establish a new weather service. They wanted the service to be a part of the government so that it could not be interrupted by the decisions and varied dispositions of private individuals.

In 1870, President Ulysses S. Grant, a former army general, approved of a congressional resolution to found a national

This statue of Joseph Henry, the first secretary of the Smithsonian Institution, stands in front of the Castle, the headquarters of the Smithsonian in Washington, D.C. (Courtesy of the Office of Architectural History and Historic Preservation, Smithsonian Institution)

weather service. Grant directed the U.S. Army to create the organization. Because weather reports were carried by telegraphy, the first national weather service was assigned to the Army's Signal Corps. There were soon 24 weather observation stations on military bases around the country.

Although the service was operated by military personnel, civilian scientists were employed as expert consultants. In 1871, a young physicist, Cleveland Abbe, was appointed to be the civilian special assistant to the commander of the Signal Corps. His appointment led to gradual improvements in weather observation, reporting, and forecasting.

Abbe was typical of the weather scientists of the time. His basic training was in astronomy and physics. After graduating from a New York college, Abbe did postgraduate work at the University of Michigan with a visiting German astronomer, F. F. E. Brunnow. He then worked as a research assistant to Professor B. A. Gould. Gould was on leave from the University of Michigan and working for the U.S. government at the U.S. Coastal Survey office in Cambridge, Massachusetts. There, Abbe became aware of the strong interdependence of weather science and the practical work of weather prediction. While preparing the publication of the *Nautical Almanac* at the survey office, he saw that scientists involved in scholarly research were

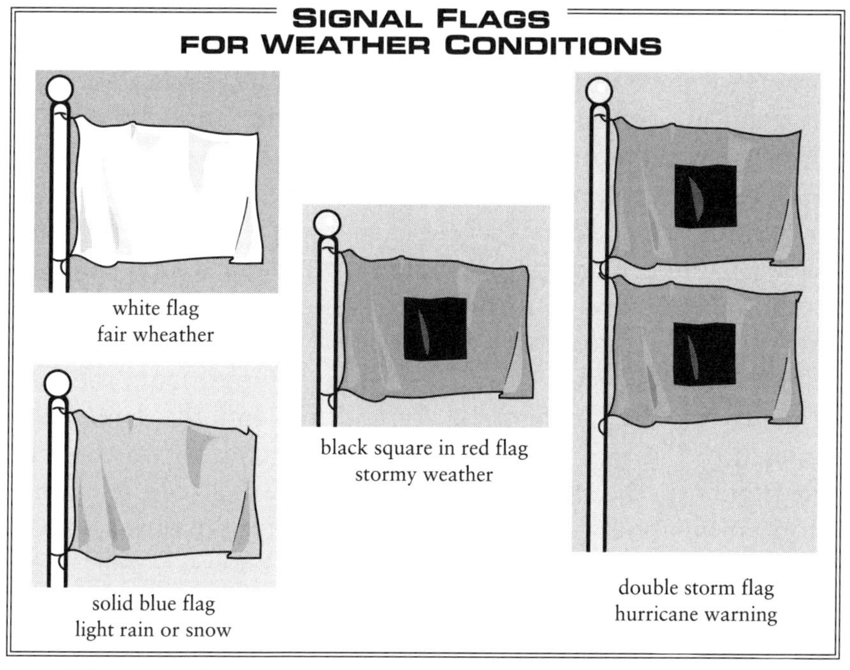

Signal flags were flown on public buildings to inform the local populace of upcoming weather conditions.

also working on projects to support sailors and fishers. He saw that scientists were gaining new ideas from these studies and from the procedures used by the weather forecasters.

After his eye-opening experiences in New England, Abbe traveled to Russia for two more years of postgraduate study in astronomy. There he found the same spirit of collaboration between scientists and weather forecasters. This collaboration was uncommon in western Europe and the eastern United States. It was a lucky accident that Abbe experienced this cooperation in two successive settings.

When he returned to the United States, Abbe attempted to establish an astronomical observatory in New York that would be similar to the Russian model. However, his efforts collapsed, and he took a position with an established observatory in Cincinnati, Ohio. After two years in Cincinnati, Abbe became a consultant to the newly formed Weather Service of the U.S. Signal Corps.

Abbe sought to expand Joseph Henry's ideas by increasing the number of stations that would record weather information. By 1878, he had established 254 observation stations, which reported to the Weather Service three times each day. The information covered temperature, barometric pressure, relative humidity, wind information, cloud cover, and precipitation. By 1888, each station made nine reports a day. Two-day weather predictions reached a new level of accuracy. The reports became more reliable in anticipating major storms that could threaten lives, damage crops, and endanger ships on the Great Lakes or near the Atlantic coast.

In 1890, the Weather Service was transferred from the U.S. Army to the Department of Agriculture and the name was changed to the U.S. Weather Bureau. Shortly afterward, Abbe and others in the new bureau recognized the need to gather information about conditions in the air above the surface of the earth. Studies of the upper layers of air are known as the "third dimension" in weather science. Meteorologists recognized that conditions in the third dimension are often different from those on the surface. For example, clouds on high can sometimes be seen to move in a different direction than surface winds.

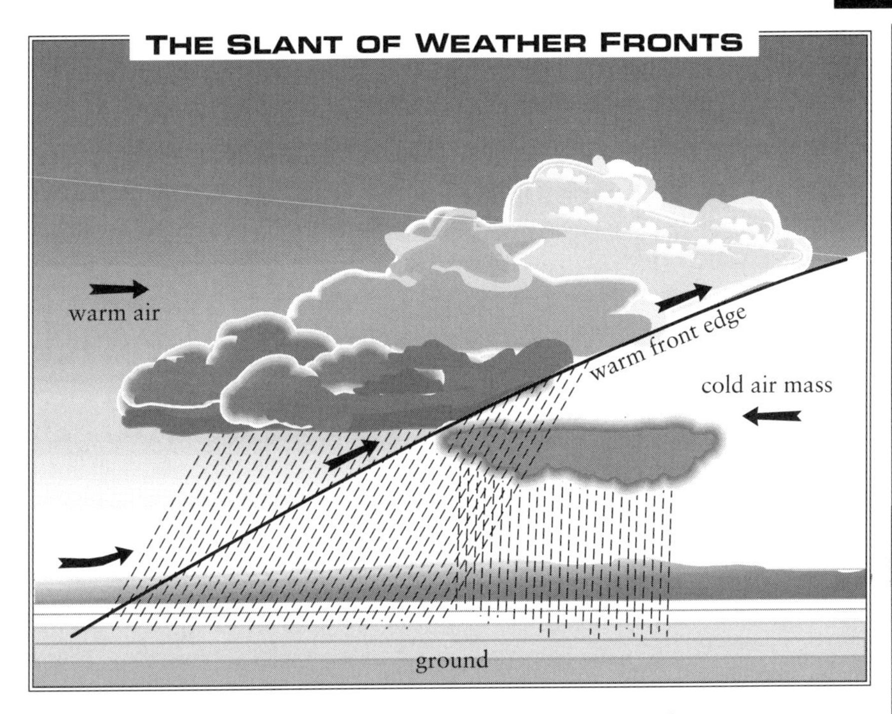

When meteorologists learned to measure temperature with instruments on kites or balloons, they realized that weather fronts were slanted rather than straight up and down.

At first, high-altitude information was gathered by weather instruments carried aloft by large box kites. However, kites could not reach altitudes high enough to satisfy the weather analysts. In 1909, scientists began investigating the use of balloons, manned aircraft, and even dirigibles to gather high-altitude information. The studies soon showed consistent weather patterns in the third dimension. For example, when masses of cold and warm air meet, the heavier, cold air tends to advance in a wedge shape under the lighter, warm air.

When the cold air is moving, the boundary between the cold and warm air is called a cold front. When a warm air mass moves to meet a stationary cold mass, the cold air stays at ground level and forces the warm air to rise along its edge. This weather condition is called a warm front and often brings rain.

These early attempts to understand climates and the weather were made before any serious attention was given to air pollution. The early work paved the way for more refined studies of weather and climate conditions. Those more refined studies then provided the foundation for showing how atmospheric conditions contributed to pollution and how industrial gases react with the natural gases of the atmosphere.

3

Modern Meteorology

During the early 1900s, European meteorologists began to investigate the movement of the air in large weather systems. One of the major figures in these studies was Vilhelm Bjerknes. Vilhelm's father, Carl Anton Bjerknes, was a physicist who taught applied mathematics at the Royal Frederik University at Christiana (now Oslo), Norway. In the late 1800s, Carl Anton's research was concerned with the natural movements of one liquid in another liquid, such as ink in water.

Vilhelm served as his father's research assistant while he continued his schooling. After completing his doctoral studies in Norway and two years of postdoctoral work in Germany, Vilhelm accepted a lecturer's position at a private college in Stockholm, Sweden. His father was pleased. There, Vilhelm would be available to edit his father's research reports and combine the material into a book. However, by the time the book was finished, new ideas in physics had outdated much of the work.

While working on his father's book, Vilhelm continued his own research. Although he did not see the connection, other scientists saw that Vilhelm's research with liquid movements was applicable to the activity of turbulent masses of air. Bjerknes was not particularly interested in these technical applications because his interests were focused on gaining recognition for his father's work. He was aware, however, that most scholars expressed more enthusiasm about the practical applications of the research than they did about the theoretical aspects.

The Physics of Weather

Gradually, the younger Bjerknes was persuaded to work less on his father's manuscripts and more on projects to improve weather predictions. At first, Bjerknes looked upon the practical side of his work as a hobby rather than serious scientific investigation. He soon realized that the Scandinavian governments funded practical research on fisheries and farming but had no great interest in theoretical physics. Bjerknes also knew that his students found good jobs working to improve weather prediction. Several of his former students soon made more money than their professor.

After 1900, weather research was becoming more important in the larger countries of central Europe. Aviation had become a respectable activity and a focus of scientific and economic interest. The Zeppelin airship was accepted as a practical means to haul cargo through the sky. Powered balloons were flown around the Eiffel Tower in Paris. The interest in human flight created a new interdependence between humankind and weather. In addition, advances in aviation allowed high-altitude observations of air temperature, air pressure, and wind. These measurements were used in new weather forecasting, which made the skies safer for flying.

By 1902, Bjerknes became preoccupied with the prospect of making meteorology into an exact science. He reasoned that he was not abandoning pure science because meteorology was a branch of physics. At the time, his argument was unrealistic because meteorology was not yet a well-structured physical science. However, the belief allowed Bjerknes to cling to his, and his father's, ideals and contemplate new avenues of research.

In the spring of 1903, his father died. While this was a sad event, it freed Bjerknes from his father's work and allowed him to apply theoretical physics to weather prediction.

On a short lecture tour in the United States in 1905, Bjerknes met R. S. Woodward, president of the newly founded Carnegie Institution in Washington, D.C. Woodward offered Bjerknes a research grant to support his studies of the atmosphere.

Bjerknes spent the next twelve years focused on standardizing methods to measure meteorological effects. For example, Bjerknes sought to standardize the measurements of barometric pressure. He also hoped to gain international agreement on techniques to chart such measurements in the form of synoptic weather maps. The term *synoptic* comes from the Greek for "seeing together" and refers to how these maps present a graphic depiction of weather information from many separate weather stations. Analysts used the information to predict future weather conditions. The early maps were difficult to use because forecasters employed a variety of symbols and notations in their graphic depictions. Bjerknes was responsible for standardizing synoptic weather maps, and his symbols are now employed around the world.

Weather predictions improved with the use of synoptic maps and because aviators began routinely reporting conditions aloft. However, many difficulties continued to exist. The number of weather stations was insufficient, and weather analysts experienced delays in receiving necessary information.

Because of his Carnegie grant, Bjerknes was able to leave his teaching job in Stockholm and return to Norway in 1907. He enjoyed living in his native country, but felt isolated from the rest of the scientific community. In 1912, after five years in Norway, Bjerknes accepted a professorship in Leipzig, Germany. The Germans had made great strides in aeronautics. They were confident that Bjerknes's research in weather science would help the field of aviation. After World War I broke out in 1914, Bjerknes remained for a time in Germany. In 1917, his government called him back to Norway to work on the practical problems of weather forecasting for farming and fishing enterprises. Norway was a neutral country during the war, and these businesses flourished because both sides in the conflict needed to import food. Consequently, the Norwegian government funded a strong program to improve weather forecasting.

Bjerknes and his colleagues set about establishing a network of observation stations and a method to speed the information to an analysis center. Many of the stations were west of Norway on ships anchored in the North Sea. Information from these western locations was invaluable because most storms came from that direction. The stations had to be situated at some distance from the west coast of Norway to allow weather analysts sufficient time to interpret and transmit vital information about approaching storms.

Transmitting the predictions continued to be a problem because the people who needed the information were scattered around the country. To improve the situation, Norway and the other Scandinavian countries invested heavily in telephone and telegraph systems during the last years of World War I and in the immediate postwar period.

Just as Vilhelm had supported his father's scientific efforts, Vilhelm's son, Jacob, entered the field of scientific meteorology. However, this partnership was far happier than the previous one. Jacob had a special genius for interpreting weather maps, and the two men made great progress in accurate predictions. Their forecasts were aided by a new source of information. During the war, the British had restricted the transmission of weather information because the German enemy might benefit from the broadcasts. After the war, weather observations from the British Isles, situated farther west than the Norwegian weather stations, were made available to the meteorologists in Norway and other European countries.

Other British Developments

In Great Britain, Lewis Richardson, a mathematical physicist, was working on ideas similar to those of Bjerknes. Richardson,

Lewis Richardson pioneered the use of mathematics to understand and predict the weather. (Courtesy of O.M. Ashford, Oxfordshire, England)

a somewhat younger man, was frustrated in his attempts to be recognized as a theorist and to obtain a faculty position at Cambridge University. He had been appointed to a series of industrial, academic, and government posts before joining the British government's Meteorological Office in 1913. Richardson used his mathematical skills to develop the first mathematical models of weather processes. The models describe, for example, the manner in which patches of relatively calm air can exist within larger storm systems. Richardson compared this air movement to the small whirlpools seen in a moving stream of water. Although the Englishman and the Norwegian did not, at first, know of each other's work, their ideas proved to be closely related.

After World War I began in 1914, the activities of the Meteorological Office were directed more and more to meet the needs of the British military services. Soon, Richardson became uncomfortable with the work. He was a pacifist because of his religious upbringing as a Quaker (a member of the Society of Friends). Richardson resigned his job and joined a Quaker ambulance unit attached to the French army. Later in his life, Richardson analyzed international conflicts and studied methods to resolve such conflicts by peaceful means.

Modern Tools

New ideas, techniques, and advanced instrumentation followed the pioneering work of Bjerknes and Richardson. Modern weather forecasting progressed to the stage where seven-day forecasts are possible. Major hurricanes are tracked from their origins in the Atlantic Ocean or the Gulf of Mexico. Affected coastal areas are given hours and even days of advance warning. Citrus growers in Florida are alerted if frost is coming. Beachgoers in New Jersey are cautioned when the bright sunshine could cause sunburn.

Weather forecasts shown on television are the result of 50 years of extraordinary advances in weather science. A bird's eye view of weather patterns—such as looking down into the eye of a hurricane—is one of the most dramatic visual presentations. Other newspaper and television images are equally spectacular. A single picture can now represent weather conditions on an entire continent. Enormous quantities of information can be summarized by the use of color graphics. Television weather forecasters can show cloud formations moving across large areas of the United States. Moreover, the movements of clouds and storms over a 24-hour period can be presented by speeded-up, graphic animation. Weather information is now understandable to everyone.

These presentations are based on three technical advances: orbital satellites, radar, and high-speed digital computers. Orbital satellites carry cameras and other instruments high above the surface of the earth. Some of these satellites circle the earth from north to south and cross over the Arctic and Antarctic regions in each orbit. Other satellites, in much higher orbits,

Many techniques are used to collect weather information. Shown here are weather ships, balloons, buoys, aircraft, satellites, radar installations, and conventional weather stations. (Courtesy of the National Aeronautics and Space Administration)

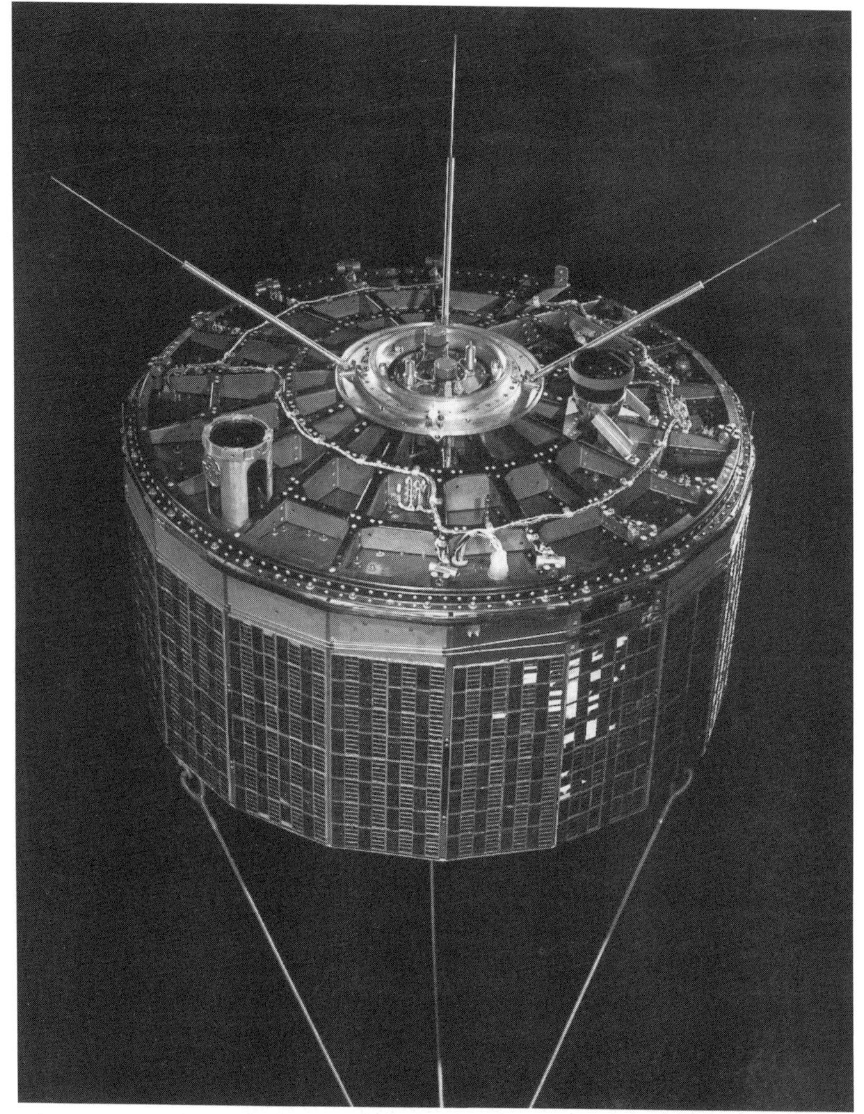

The TIROS VII *weather satellite has two television cameras to detect weather patterns in the atmosphere near the surface of the earth.* (Courtesy of the National Aeronautics and Space Administration)

remain permanently stationed over one area. These are called geostationary satellites. Both types of satellites take digital photographs of the weather activity below them. The information is sent to ground stations by special electronic communication links. These pictures greatly expand the information supplied by weather stations on the ground and provide new perspectives for weather analysts.

Radar provides another major advance in the assembly of information. Radar sends out electronic signals that are reflected by objects in the air or on the ground. The reflected signal pattern is detected by an antenna and relayed to a special television tube. The patterns are interpreted by a professional meteorologist. Early radars could detect only solid objects, but the newer radars obtain reflections from falling rain and even from clouds. The best ground-based radars can discern small areas of intense turbulence that could easily escape detection by satellites. These technological advances help meteorologists to record thousands of observations over extended periods of time.

Solar energy provides electric power to this computerized weather station at Stratford Shoals, Connecticut. The station automatically transmits weather information to a weather analysis station. (Courtesy of the National Aeronautics and Space Administration)

Modern computers allow the integration of such weather information into a single, highly meaningful representation. The computers have been programmed to perform specific mathematical operations on the data supplied by satellites, radars, and other standard weather instruments. The computer programs are based partly on the mathematical models first conceived by Bjerknes and Richardson. These computer models are vastly improved. For example, a computer can be programmed to draw weather maps.

Without these capabilities for understanding the hourly movement of air masses and the long-term cycles of climate, it would not be possible to grasp the effects of air pollution.

4

Weather Modification

For thousands of years, humans have tried to influence the weather by songs, dances, and prayers to their gods. Shamans have presided over rituals to ensure good growing seasons. Today, people who live in rural societies are more attuned to the weather than those who live in cities. When news broadcasters announce that farmers are experiencing a drought, city dwellers merely feel they are enjoying another fine, dry, sunny day.

That people discuss the weather and complain about it, but do nothing to change it is not exactly true. People have always controlled some aspects of their weather experiences. They have taken shelter in caves, tents, grass huts, and brick buildings from rain, snow, and hail. Even in primitive structures, a window, door, or other opening allows air circulation. A system of ventilation—no matter now crude—is a form of weather technology.

Microclimates

When humans wear clothing, they create a microclimate in the environment nearest their bodies. When they are inside a building, they experience two microclimates. The environment of the

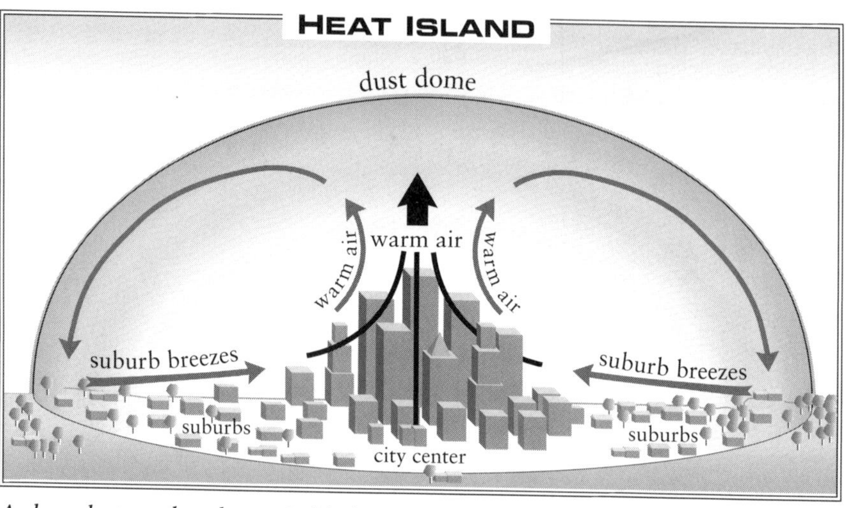

A densely populated area is likely to produce a heat island. Consequently, cities have a different climate from the surrounding countryside.

interior of the building differs from that of the exterior, and the conditions inside their garments differ from those outside the clothing.

Farmers are experts in establishing microclimates. In addition to providing a sheltered environment for their livestock, they shield their crops from adverse weather in several different ways. They provide wind breaks of dense shrubbery or trees to separate and protect their fields. They supply synthetic rain by the use of large sprinkler systems. In the citrus groves of California, Texas, and Florida, growers prevent frost from damaging the fruit by using water mists and artificial smoke to hold radiant heat near the ground.

Nature also produces microclimates. Coastal lands are part of a microclimate because the nearby water serves to moderate the weather. On the flat prairies of the Midwest, any rise in the contour of the land serves as a wind break, and conditions differ from one side of the hill to the other. However, some of the most dramatic microclimates are produced by human technology.

Every town and village has a climate that differs slightly from that of the surrounding countryside. In big cities, the differences in climate between city and nearby country can be fairly large. Cities generate what are called heat islands. On hot summer days, asphalt streets and brick buildings absorb and retain heat from the sun. At night, this heat is slowly radiated into the cooler air. Air conditioners transfer heat from the inside of a room or a building to the outside. All of a city's industrial processes generate surplus heat. In winter, buildings leak warm air from their heating systems and brick buildings absorb and then radiate any heat from the sun. Snow seldom affects the core of a big northern city such as New York. The heat rising from the center of the city can melt the snow before or soon after it hits the ground.

Making Rain

The most important large-scale weather modification is the control of rainfall. In 1891, the secretary of agriculture—the cabinet member then in charge of the U.S. Weather Bureau—attempted to produce rain. He instructed his weather technicians to arm balloons with explosives and send them aloft. He hoped to generate rainfall by setting off the explosives when the balloons reached the clouds. C. W. Post, the founder of a major breakfast cereal company, tried similar experiments in 1911 and 1912. Fortunately, neither experiment was successful. Otherwise, explosions might have become a regular—and perhaps dangerous—response to every dry spell.

The first modern approach to rainmaking began in the 1930s. These early efforts were based on the way condensation occurs when warm, moisture-laden air is cooled quickly. The effects of condensation can be seen on a hot, muggy day when droplets of water form on a pitcher of iced lemonade. Scientists hoped that the same principle would apply if shaved ice were dropped into a moisture-laden cloud. The results were disappointing because no single aircraft could lift enough ice to make any impact on the clouds. Using a fleet of large aircraft would be too costly. In any case, shaved ice is an ineffective way of cooling the interior of a cloud. Indeed, the cloud's interior may contain cold air that is already below the freezing point of water.

Just before World War II, the development of high-altitude aircraft increased the knowledge of cloud structures. Scientists realized that the fine mist that makes up the body of the cloud can be very cold. German meteorologist Walter Findeisen theorized that rain occurs when the water in the cold mist condenses on solid particles and then freezes around them. When the frozen particles move through the cloud, they capture more water and grow larger and heavier. When they get heavy enough, they fall into warmer air, melt, and descend as rain. Findeisen decided that, in nature, the original starting particles were probably tiny ice crystals. In the language of weather science the crystals are called condensation nuclei—the nucleus or core on which the water condenses.

Findeisen wondered whether artificial crystals could be substituted for the naturally formed ones. He tried to seed clouds with condensation nuclei made of quartz and other materials, but the results were disappointing.

After the war, Vincent Schaefer and Irving Langmuir, a Nobel Prize winner in chemistry, began a program of meteorological research under the auspices of the General Electric Company. Langmuir had become interested in clouds from working on smoke-screen generators during the war. Schaefer was initially interested in the wonders of snowflakes and their near-perfect six-sided symmetry. Schaefer wanted to photograph individual snowflakes. He tried repeatedly to create the flakes from cold, moisture-laden air inside a refrigerator lined with black cloth. Nothing worked. One day, he forgot to close the refrigerator door when he took his lunch break. By the time he discovered his mistake, the refrigerator had warmed. Schaefer decided to use dry ice—frozen carbon dioxide—to correct the problem as quickly as possible. Dry ice is much colder than water ice, and it cooled the temperature of the refrigerator far below the freezing point of water. When Schaefer reopened the door and peered inside, his moist breath froze and became tiny ice crystals.

Quite by accident, he had discovered that the extreme cold produced by dry ice could generate condensation nuclei from moist air.

Advanced Techniques

Experiments soon moved from the laboratory to the real world. Meteorologists demonstrated that condensation nuclei were produced when dry ice was released into clouds from an aircraft. These experiments verified Walter Findeisen's theory that the water carried by clouds condenses on crystals—either natural or artificial. Water then freezes on the nuclei, and the crystals grow larger and heavier, fall toward the earth, and depending on temperature conditions, descend as rain, snow, or hail.

This weather-modifying technique was found to change the appearance of the rain clouds. Pilots conducting the tests reported that holes appeared in the clouds as soon as they were seeded with dry ice.

Further study showed that silver iodide crystals also could serve as condensation nuclei. When vaporized by intense heat, solid silver iodide breaks up into billions of tiny particles. The particles are almost weightless and resemble smoke or fog. The "smoke" of the vaporized silver iodide is so light that cloud seeding can be initiated from the ground. Wind currents can carry the "smoke" aloft and into the path of a prospective rain cloud. However, the most common seeding technique employs specially equipped aircraft to carry the silver iodide. No matter how the particles reach the clouds, each tiny sliver of silver iodide becomes the center of an ice crystal that eventually falls on the earth as a form of precipitation.

Since the end of World War II, large-scale cloud seeding has become a commercial practice. In the United States, private companies are employed by county and state governments. The companies claim that they can increase rainfall by about 15% in a specific area during a given period of time. They also believe that cloud seeding techniques can be used to prevent hail storms and to disperse fog. However, such claims are difficult to verify. Some of these weather changes may be the result of natural variations rather than human attempts at modification.

Some Reservations

Manipulating the weather can cause problems. When moisture-laden clouds are seeded to drop rain in a certain area, the clouds will be out of rain when they travel onward. The people in a dry locale might accuse the rainmakers of stealing their rain. If a group of farmers hires a company to cause rain on the same day as a local golf tournament, the golfers might be angry. Such cases have actually gone to court.

Government research on weather modification takes place at the National Center for Atmospheric Research in Boulder, Colorado. Scientists and technicians from this center test the most advanced techniques. For example, under an international agreement, special flares are being tested in the desert areas of Mexico. These flares generate microscopic particles that act as condensation nuclei. Aircraft fly under thunderclouds and launch the flares upward into the clouds. The results have been encouraging. In some cases, this technique has increased rainfall about 30% over the normal amount.

However, many weather scientists argue that these conclusions may be based on wishful thinking. Specialists with no financial interest in weather modification suggest that seeding techniques produce an increase in rainfall of only 2%–3% rather than 15%. They view these techniques as feeble, low-energy procedures that are too weak to upset the normal processes of a large rain cloud.

5

Local Air Pollution

Some air pollution comes from natural sources. Volcanoes can produce large quantities of toxic materials. Hot springs create modest amounts of noxious gases that contain the element sulfur. Forest fires caused by lightning generate air pollution. These fires produce some carbon dioxide, small amounts of toxic carbon monoxide, and oxides of nitrogen. Human activities, however, produce much larger amounts of these and other pollutants.

Early Cases

Fire is the most basic process underlying advanced technology. Fire is also the greatest contributor to air pollution. Since prehistoric times, fires kept people warm in winter, cooked food, lighted dark places, kept predators away, and hardened clay vessels. The earliest known fired clay pots were found in the Near East and dated around 7000 B.C. Fired vessels were used for cooking and for transporting and storing foods, liquids, or commodities. Their usefulness led to the technology of ceramics.

Fire is also used in the production of an important building material. For countless centuries, bricks were shaped from mud or clay and hardened in the sun. Around 3000 B.C., people in the Near East began to produce fire-hardened bricks. The molded clay was heated to a high temperature in a kiln or oven. Fire-hardened brick is a long-lasting and desirable building material.

The smelting or purification of metals is another ancient industry that requires the use of high temperatures. Small amounts of pure gold are found in riverbeds. Fire is not always necessary to purify gold. However, copper ore always contains sulfur or other foreign materials. Therefore, the ore must be purified by the use of intense heat. Archaeologists believe that copper was probably the first metal to be smelted. This industry may have begun in the Near East as early as 5500 B.C. and in Europe around 3500 B.C. Metals were smelted—melted from their ores—in primitive furnaces.

Usually, copper ore is found in the form of copper sulfate. When the ore is heated to the proper temperature, the sulfur is driven off as sulfur dioxide gas (SO_2) leaving the pure melted copper.

Although objects made from copper might be very beautiful, the sulfurous gases released by copper smelting are very unpleasant. A compound of sulfur and hydrogen has the same smell as rotten eggs and is disagreeable in even very small amounts. Gaseous compounds of oxygen and sulfur are far more dangerous because they can change to sulfuric acid in moist air. In those ancient times, copper smelting was a small-scale industry, but it may have been responsible for the first hazardous industrial pollutants. In fact, all ancient industries that required the use of fire were adding pollutants to the air.

In more recent times, the Industrial Revolution in Britain and northern Europe greatly increased metal smelting operations. Smelting and other industries caused a dramatic increase in the quantity of pollutants sent into the atmosphere. Indeed, the amount of pollution has doubled every 10 years from the early 1800s to modern times.

Specific Effects

The Industrial Revolution took place in the years between 1750 and 1850. At that time, factories using newly developed complex machinery were hiring vast numbers of workers. People moved from farm communities to large cities to be near their jobs. The overcrowded cities were unable to cope with the new arrivals, and all areas of life were adversely affected. The industries were not regulated and had few rules regarding child labor, pollution, or the health and safety of workers. Unsafe gases and poisonous chemicals routinely escaped from the factories' smokestacks and pipes.

By the early 1800s, factories making lye and other caustic materials were sending especially toxic air pollutants into the atmosphere. These factories released hydrochloric acid gas in their smoke. It was soon apparent that the discharges were extremely harmful to the vegetation and buildings downwind of the factories. (The health of the workers inside the factories was not an issue.)

The British government eventually passed the Works Regulation Act of 1863. The act directed factory owners to remove hydrochloric acid from the smoke by passing it through large charcoal filters. This technique removed 95% of the acid. The owners found that the recovered hydrochloric acid was a valuable industrial product. Therefore, the law achieved a double benefit. The environment was protected, and the owners made money from the sale of the former waste material.

In the United States, early pollution problems were caused by large copper-, zinc-, and iron-smelting mills. Today, past environmental destruction is still visible in places where these metals were mined and processed over long periods of time.

In the mid- to late 1800s, steelmaking became a major industry in the United States. By 1900, the work was concentrated in locations such as Pittsburgh, Pennsylvania. Because steel production requires large amounts of both iron ore—the raw material of steel—and coal, these mills were built near centers of river and rail transportation. The iron ore was shipped in from

Minnesota, and the coal was delivered from nearby coal mines. Although iron ore contains little sulfur, the coal used to heat the ore often has significant amounts of that element. The sulfurous gases sent into the atmosphere were devastating to growing plants. Many local farmers were ruined when the gases killed their crops.

During the mid-1800s, when industrialization was expanding rapidly through the United States, the government did little to regulate the industries. Protection from property damage and health hazards depended on local regulations and law courts. Politicians and judges knew that any attempt to regulate industrial activities could mean a loss of employment for the local citizens. The methods required to safeguard the atmosphere were costly, and when confronted, some owners simply moved their factories to different areas. Workers were given a choice. They could accept discomfort and health hazards, or they could lose their jobs. Similar situations still occur.

At that time, a lawsuit against the factory owners was the only way farmers could obtain compensation for the damage to their crops or livestock. If the farmer could prove that the factory was directly responsible for the loss, the local legal system would determine how much the factory owners should pay. However, few farmers had the money for a lawsuit. The offending industrial firm, on the other hand, usually could afford good defense lawyers. Not many individuals wanted to go into court unless they had strong evidence to support their cases.

As recently as the late 1940s, farmers in Tennessee complained that their crops and animals were being damaged by exposure to fluorine gas. The gas was being released during the purification of aluminum in some areas and during the production of phosphate fertilizers in others. The farmers' complaints resulted in a series of field studies conducted by the U.S. Department of Agriculture. The findings from the field studies showed a correlation between the level of production at the industrial plants and the amount of damage to crops and livestock. In 1951, the department sponsored a meeting of the interested parties in Washington, D.C. However, a direct causal link could not be proven by the correlation. The exchange of views between the farmers and the industrialists did little to resolve the problem. The farmers received no compensation because their claims were never settled.

The injustice of this situation eventually brought other federal agencies into the dispute. However, these officials were also reluctant to restrict industrial operations that provided employment and valuable products. Today, national and local governments seek a balance between the well-being of some individuals and the employment opportunities of others.

Smog as Local Pollution

Smog comes in two varieties: sulfurous and photochemical. Sulfurous smog began to appear at the height of the Industrial Revolution in the late 1800s. No one knows when photochemical smog first appeared in the atmosphere. However, scientists became concerned about this pollutant when automobile traffic expanded after World War II.

Smog is found mainly in or near large cities. Local weather conditions, local geography, and emissions from vehicles and factories all contribute to this dangerous form of pollution. Smog becomes most severe when the air is very calm and no wind disperses the polluting materials. Temperature inversions are particularly troublesome. Inversions occur when a warm mass of air sits atop a cooler, heavier air mass. If the cool mass lies in a valley or large low-lying area, the heavier air may not move for several days. Therefore, this stagnant air collects huge amounts of pollutants from industrial or automotive emissions or both. Los Angeles is famous for both terrible smog and long-lasting temperature inversions.

SULFUROUS SMOG

In 1905, an English public health official coined the word *smog* to describe a mixture of smoke and fog. In those times, the smoke from household coal fires contained significant amounts of sulfur dioxide and fine particles of sootlike materials. The water vapor (H_2O) in the foggy portion of the smog combined with the sulfur dioxide (SO_2) in the smoke to form sulfuric acid (H_2SO_4). This made the smog a serious health hazard.

Six years after coining the term, the same public health official announced that smog had claimed many lives in Edinburgh and Glasgow, Scotland. In 1930, sulfurous smog was thought to be responsible for 60 deaths in an industrial valley of Belgium. Twenty people died and hundreds were made ill in 1949 by a heavy smog in Donora, Pennsylvania. Four thousand fatalities were assigned to a four-day "killer" smog that struck London, England, in 1952.

Most of the victims were elderly people who suffered from chronic lung or heart problems. Those who wished to minimize the problem said that the smog may have shortened their lives but was not the sole cause of death. However, the evidence from long-term studies of air pollution strongly suggests that permanent damage to the lungs and circulatory systems builds up over time. This is true even when the pollution levels are relatively low. In any case, the evidence was sufficient to lead national governments to begin to restrict industrial emissions containing sulfur compounds.

For many years, burning coal was used to heat homes as well as generate electric power or smelt metal. At present, home heating is a minor source of sulfurous smoke. Electric power plants and other factories have continued to generate sulfur dioxide and other hazardous chemicals. Gradually, these companies realized that they must reduce the contamination caused by their pollutants. They developed better methods of combustion and waste disposal. The improved combustion yielded more heat from the same amount of fuel. The cleanup efforts produced additional benefits in the form of reclaimed, marketable materials. Consequently, improving the quality of air by reducing harmful emissions has a positive effect for both industry and the local communities.

PHOTOCHEMICAL SMOG

At present, photochemical smog is the more dangerous of the two smog hazards. Most photochemical smog is generated by materials released from automobile exhausts. The most prevalent pollutants include gaseous oxides of nitrogen, vapors from unburned gasoline, and fine particles of carbon-based compounds. When sunlight strikes these gases, ozone is formed by a complicated series of chemical reactions. Each molecule of ozone is composed of three atoms of oxygen (O_3). Therefore, ozone differs from life-giving oxygen molecules, which have two atoms of oxygen (O_2).

Ozone can cause stinging eyes and a raspy cough in anyone unlucky enough to be exposed to the gas. People who have chronic lung conditions or asthma are very vulnerable to ozone exposure. Such people are warned to limit their activities on smoggy days so that the bad effects will be minimized.

Photochemical smog becomes most dangerous when three conditions prevail. The first is heavy automobile traffic. The second is bright sunlight. The third is a stagnant air mass. Sometimes the air within a densely populated urban area generates a heat island. In such circumstances, the air tends to rise in the area of closely packed brick buildings and black-topped streets. The bricks and the tarred streets both absorb heat from the sun and release it when the air cools. The warmed lighter air

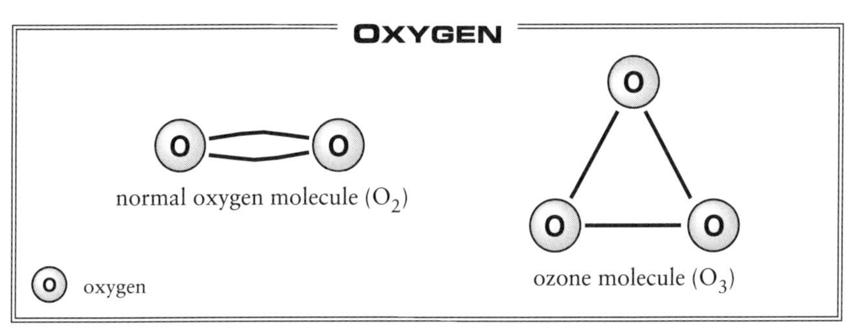

The common oxygen molecule has a double bond between its two oxygen atoms. Ozone has a single bond between each oxygen atom, which makes it much more reactive than the ordinary oxygen molecule.

ascends over the area, cools as it reaches a higher altitude, descends over the suburbs, and finally reenters the city. In other words, the same air circulates and recirculates within the city's boundary. During this time, pollution from automobile emissions and factories accumulates and becomes more and more concentrated.

A common factor in the generation of pollution is a temperature inversion, such as experienced by the city of Los Angeles, California, and other cities built in valleys or surrounded by hills. In the late 1940s and the 1950s, photochemical smog reached dangerous levels in Los Angeles. That city has very bright sunlight most of the year and unusually heavy automobile traffic. It is built on a very low-lying plain surrounded on three sides by mountain ranges. In short, Los Angeles provides all the critical conditions for smog production.

In the 1950s, the smog situation in Los Angeles was becoming intolerable. Smog alerts were broadcast on radio and television almost daily. Elderly people and those with lung problems were warned to restrict their activities. Infants were also vulnerable. Any active sport that caused rapid breathing was becoming risky. By 1954, experts recognized that exhaust products from automobiles were the main source of photochemical smog.

Public officials and representatives of citizens' groups in California complained to their state and national legislators. Hearings were held at the state capital and in Washington, D.C. The National Academy of Sciences in Washington was asked to form a panel to review the scientific and technical aspects of the problem. During the 1960s and early 1970s, the National Research Council (part of the National Academy) produced a series of increasingly negative reports.

Some form of regulation was obviously needed. However, government officials—particularly, those in the newly established Environmental Protection Agency (EPA)—were unsure of a solution to the problem. They considered restricting the use of automobiles to certain days of the week. (That method is now employed in Athens, Greece, where the climate and other factors are similar to those in Los Angeles. It has not been very successful.)

After much thought, legislators in Congress and officials at the EPA sought a technical solution to the problem. They reasoned that smog would be greatly reduced if pollution could be stopped at the source. The experts determined to find a way to reduce or eliminate the pollutants found in automobile exhaust fumes. These pollutants include gaseous oxides of nitrogen and other impurities. Automobile exhaust also contains carbon monoxide (CO), which is toxic by itself. A car going one mile (about 1.6 kilometers) generates almost one ounce of carbon monoxide (28 grams). To dilute this toxic gas to a safe level requires about 2 million quarts of air (1.9 million liters). In other words, the dilution factor is 64,000,000 to 1.

For 20 years, scientists and engineers affiliated with universities and private industry worked on the problem of removing contaminants from auto exhausts. Progress was slow because little basic research had been done before the crisis. In fact, most previous research on the chemistry of nitrogen was directed toward development of nitrogenous fertilizers. That procedure included the capture of pure nitrogen gas and transforming it into nitrogen oxides. Such chemical reactions had exactly the opposite effect compared to reducing nitrogen oxides to pure nitrogen, which is not a pollutant.

Finally, the scientists and engineers provided a solution to the problem of smog control. They designed a device known as a catalytic converter. The converter is a small canister connected to the exhaust pipe of a vehicle. It contains a series of ceramic screens coated with platinum and other rare metals. Chemicals in the hot gases generated by a gasoline engine pass through the canister and are broken down into relatively harmless molecules. In a complicated sequence, each metal acts upon a different pollutant. Nitrogen oxide becomes molecules of pure oxygen and pure nitrogen. Unburned gasoline, carbon monoxide, and particles of carbon-based substances are broken down into water vapor and carbon dioxide. The carbon dioxide gas is the only contaminant that remains. The problem of California's smog had been moderated.

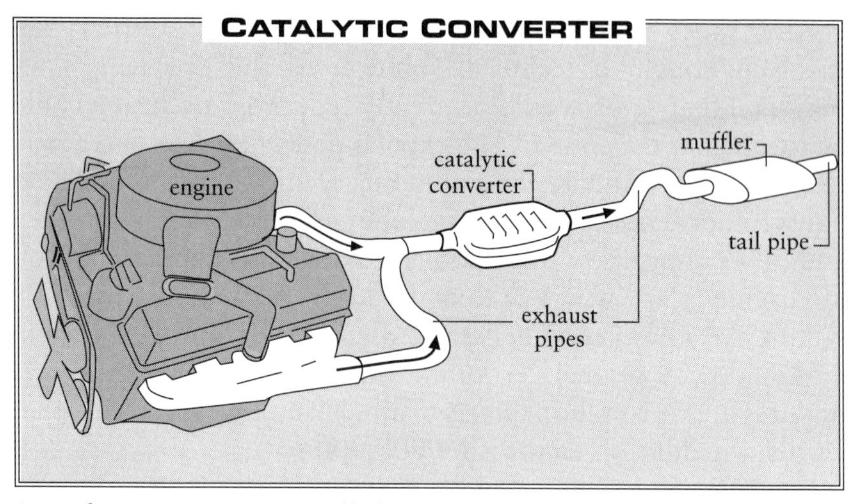

A catalytic converter is installed in a car's exhaust system in front of the muffler. It converts exhaust gases to pure nitrogen, water, and carbon dioxide.

For a few years, people who lived in other areas of the United States had been rather indifferent to the smog in Los Angeles. They considered themselves better situated than the poor smog-plagued Angelenos. However, it soon became apparent that many other cities—some of modest size—were also susceptible to the problem.

By 1975, photochemical smog was beginning to appear in many cities. That same year, officials of the Environmental Protection Agency (EPA) put forth a regulation to improve the situation. They decreed that the newly designed catalytic converter must be installed in most cars manufactured in 1976. The following year, the requirement applied to every new automobile and truck manufactured in the United States.

The timing was not good for several reasons. The converters were still rather crude and not fully efficient. Also, citizens of the United States were experiencing a serious gasoline shortage. A petroleum embargo had resulted from political developments in the Near East, and little oil was reaching refineries in the United States. People were concerned about trying any new device that might increase the need for gasoline.

In addition to the other problems, many groups in the United States were opposed to the required installation of catalytic converters. Auto manufacturers saw the addition of this device as interfering with attempts to hold down their costs. Some people claimed that the catalytic converters reduced the efficiency of the car's engine and increased fuel consumption. These claims were generally untrue but widely believed. Many new car buyers broke the law by paying a mechanic to remove the device. The owners believed that they were conserving precious gasoline by their illegal act.

In their haste to solve a dangerous problem, regulators had mandated the use of catalytic converters before they were fully perfected. Converters installed in the first few years of the EPA ruling were worn out before the car became outmoded. It was not until 1980 that the catalytic action could continue to perform for 50,000 miles (80,000 kilometers). Today, the technical standard of converter performance requires at least 100,000 miles (160,000 kilometers) of service.

ENFORCEMENT

The functioning of catalytic converters can be monitored by regular inspections. Since the early 1980s, 28 states and the District of Columbia have required periodic testing for automobiles registered in high-traffic areas. The exhaust gases are sampled, and the amount of contaminants is determined. If the car does not pass the test, the converter must be replaced at the owner's expense.

Fuel Complications

In 1921, a chemist, Thomas Midgley, Jr., discovered that the addition of lead allowed the gasoline used in automobile engines to burn more evenly. Carbon-based molecules containing lead were first introduced as gasoline additives in the mid-1920s.

Although the lead added to the engine's efficiency, it later proved to be one of the contaminating materials in automobile exhausts.

In the late 1940s, medical research indicated that lead ingested in food and drinking water was a danger to health. Lead—found in particles of old paint and other substances—caused retardation in children, problems of the central nervous system, and impairment of kidney function. In the 1950s and 1960s, public health officials began a campaign to reduce the amount of lead taken into the human body.

As the number of automobiles increased, lead from their exhaust was also found to be a significant source of lead poisoning. In addition, scientists discovered that the lead that affected human well-being destroyed the catalytic action of converters. Therefore, the legally mandated use of converters brought an end to the use of leaded gasoline. Petroleum companies were forced to provide lead-free gasoline to preserve the usefulness of catalytic converters. Thus, the introduction of catalytic converters has had two beneficial results. Lead-free gasoline eliminates lead emissions, and catalytic action greatly reduces the amount of smog-producing gas. Other nonmetallic chemicals are now used to perform the same functions as lead in gasoline.

Catalytic converters operate at a very high temperature. The device remains hot for several minutes after a car is parked and the motor is turned off. Unfortunately, in those few minutes, the hot converter—located on the underside of the car—could start a fire in a pile of leaves, for instance. In 1996, engineers at the National Renewable Energy Laboratory and at Benteler Industries, Inc. have developed a special insulating cap that fits over the catalytic converter. The cap shields the surrounding area from the intense heat generated by the converter. It also retains the heat so that when the driver restarts the engine after a brief interval, the still-hot converter is ready to perform. Because the catalytic converter must be hot to remove pollutants, the insulating cap helps improve the converter's performance while preventing accidental fires.

Smog Remains

Unfortunately, the smog problem is still not totally eliminated. The catalytic converter can remove only 90% of the harmful emissions. Also, the number of automobiles on the highways of the United States continues to grow. Consequently, after a temporary decline, the amount of nitrogen oxides and unburned fuel particles is increasing again. Several conditions contribute to this resurgence. Engineers have succeeded in designing cars that last longer. Even in the newest models, car engines and catalytic systems become less efficient over time. In general, cars more than 10 years old are the biggest offenders. Some environmental research organizations estimate that 10% of the automobiles generate more than 50% of the pollution. That 10% includes most of the older cars.

California and other states have considered a program of buying older cars from their owners. However, forcing citizens to sell their older cars is not politically sound. Many citizens believe that this policy would be unwarranted government interference in their lives.

Additional technical solutions are on the horizon. After early enthusiasm and many later disappointments, progress is being made to fabricate a car that runs on electricity. Since electric power generation also produces some pollution, the electric car is not a perfect answer. If car engines are redesigned, the use of alternate fuels such as natural gas (methane) could improve the situation. Natural gas produces much less pollution-causing nitrogen oxides than other fuels. However, with today's technologies, neither electricity nor natural gas would offer a complete solution to the problem.

One of the most appealing alternative fuels is pure hydrogen. It is very efficient and generates only water vapor as an emission. However, pure hydrogen can be costly to produce and its use will require changes in engine design.

Nontechnical Adaptations

For several decades, people have tried various programs to reduce automobile use. Car pooling is one such program. Many states, notably Virginia, have reserved express lanes on major highways for vehicles that carry more than one person. This practice provides an incentive to car pool because driving on the uncongested express lanes is faster and less stressful. As an added benefit, each member of the car pool saves money on gasoline and car repairs. Also, many people like the companionship.

Other programs sponsored by local or regional governments include the expansion or restoration of mass transit systems such as buses and subways. In relatively recent times, subways have been built in the metropolitan areas of San Francisco, California, and Washington, D.C. Modern subways are an excellent solution to traffic and pollution problems. The amount of pollution produced per passenger trip is far lower than that generated by an automobile. Although this type of transportation has a strong appeal to planners and many citizens, subway ridership has not grown to the expected level.

In Europe and Asia, the bicycle has long been an important mode of transportation. Urban planners in the United States have tried various methods—such as specially marked bike lanes—to popularize bicycling in the United States. However, except for purely recreational purposes, the use of bicycles for transportation has been limited mainly to young people.

In spite of some success in the programs to reduce photochemical smog, this type of pollution remains a major problem.

6

Acid Rain

For many years, emissions from smelting and other industrial facilities have caused damage to nearby property. Fortunately, restrictions on industrial practices have partially solved the problem. However, the danger of pollution remains.

The combustion, or burning, of fossil fuels—coal and oil—releases gases that can generate acid rain. This threat occurs when the gases of sulfur dioxide (SO_2) and nitrogen oxides combine with oxygen and moisture in the atmosphere. The molecules of nitrogen oxides then form nitric acid (HNO_3) and the molecules of sulfur dioxide form sulfuric acid (H_2SO_4). To complete the process, the acid is dissolved in drops of rain and eventually falls to the ground. Forests, fields, gardens, and aquatic plants are harmed by these chemicals.

Sulfur emissions are the most important cause of acid rain. In the United States, electric utility companies produce 70% of the airborne sulfur dioxide. In 1980, the smokestacks of electric power plants in the United States sent 11 million tons of sulfur dioxide into the earth's atmosphere.

Nitrogen oxide—the other chemical responsible for acid rain—results from the combustion of oil, gasoline, and jet fuel. Although both chemicals are harmful, scientists have found that the acid rain attributable to nitrogen oxide tends to be more local than that caused by sulfurous gases.

A major difference between the effects of sulfurous smog and acid rain is the size of the affected area. The aftermath of smog is a local problem, while acid rain damage can extend over a very large area. Moreover, the areas damaged by acid rain can be found hundreds of miles from the source of pollution. The contaminated smoke is sent high into the air by the heat of combustion and the height of the smokestacks. Strong winds can carry the damaging gases for long distances before they fall to the ground.

In North America, the chemicals that form acid rain are carried from west to east by the prevailing winds. Wind-blown sulfurous smoke from electric power companies in midwestern states travels hundreds of miles before descending as acid rain in the northeast. Areas along much of the Atlantic seacoast have been harmed by this pollution. Officials of Canada's eastern provinces and the New England states have reason to be upset by the emissions from these power plants.

Although acid rain poses no direct health hazard to humans, it attacks paint on buildings and erodes limestone structures and sculptures. It reduces food crop production and can kill or dwarf pine trees. In higher, more exposed locations, spruce trees and fir trees in New York, Vermont, and New Hampshire have been destroyed. Acid rain has weakened trees in the forests of Tennessee, Virginia, and North Carolina. The damaged trees become vulnerable to insects and soon die.

Acid rain can sterilize lakes by increasing the acid content of lake water. The acid kills the small plants that feed the small fish that are eaten by large fish. Sport fishing is an important part of the tourist industry in Canada and the northeastern United States. When lakes become barren, the tourist business declines.

In the early 1980s, the problem of acid rain strained relations between Canada and the United States. Canadian political officials wanted the United States to regulate sulfur dioxide emissions or pay compensation to Canadian enterprises—or both. The discussions became heated, but no solution was forthcoming. Negotiators for the U.S. government saw that restraints on electric power generation would have negative political and economic consequences in the midwestern states. However, when states all along the east coast began to complain about the situation, the U.S. government finally was forced to take action. At first, utility companies were encouraged to reduce emissions on a voluntary basis. Then new, stronger regulations were added to existing laws. In 1983, the Federal Air Quality Act was passed, and in 1990 it was revised to further the control of emissions.

Under the provisions of this act, regulators were required to study the records of each utility company. From this informat-lon, they calculated the amount of sulfur dioxide that had been generated by each factory in the preceding years. Because sulfur dioxide gas is the most frequent pollutant, this figure determines the tonnage of pollutants, or allowances, allotted to each factory or power plant.

Each allowance is equal to one ton of sulfur dioxide emissions. If a company had a history of putting 3,000 tons of sulfur dioxide into the air every year, it was allotted 3,000 allowances. No more allowances would be issued even if the utility company expanded its power production level. Therefore, under the new plan, the company could produce 3,000 tons of pollutants without penalty. If it generated less than its allotted amount, it could sell the extra allowances to another company. On the other hand, if the company generated more sulfur dioxide than allowed, it is required to buy additional allowances from some other utility company. If it does not, the government applies a penalty of $2,000 per ton of unallotted pollutants.

The purpose of the Federal Air Quality Act of 1990 was to prevent an increase in pollution at the same time that the demand for electricity expanded. However, officials hoped that the effects of the law would gradually decrease the level of pollution. To the surprise of many environmentalists, this scheme has worked reasonably well. The rate of sulfur dioxide emission in 1980 was about 11 million tons per year. By 1995, the emission level had been reduced to a bit more than 5 million tons. The pollution level had been cut in half.

While many New Englanders and Canadians are still dissatisfied, the Environmental Protection Agency is pleased with the record. They point to the increasing demand for electricity between 1980 and 1995. In earlier years, this increase would have resulted in a higher pollution level. The Air Quality Act has brought about a cleaner environment at the same time that most utilities companies enjoyed high employment and economic prosperity.

Technical Responses

The new rules and regulations caused the managers of utility companies to investigate ways of reducing pollution from their power plants. Scientists sought methods to decrease the amount of contamination caused by coal smoke. Therefore, they set out to remove some of the sulfur products that are present in coal.

Two types of coal are mined in the United States. Anthracite, or hard, coal has a high carbon content, a low sulfur content, few impurities, and burns with a clean flame. Bituminous, or soft, coal contains many complicated carbon compounds, more sulfur, many impurities, and burns with a smoky flame that generates many pollutants. The hard coals are therefore superior to the soft coals for pollution control. However, the hard coals are more expensive. Consequently, special techniques for "washing" bituminous coal have been developed.

When soft coal is mined, relatively small quantities of the surrounding rock, sand, and gravel are extracted with the ore. All of these materials contain unwanted sulfur compounds and various other impurities. In order to separate the coal from the unwanted materials, the newly mined coal is dumped on conveyor belts, and strong jets of water or air are directed at the mixture of coal and rock. Because coal is lighter than rock, the force of the water or air moves the pieces of coal away from the heavier rock. Although the washing process does not change the sulfur content of the coal, it physically separates the coal from the mix of heavier, sulfur-rich rock and gravel. These foreign materials are discarded, and the coal is made ready for shipping.

Although washing coal reduces the amount of pollutants, this technique does not eliminate the pollution problem. To help correct the situation, engineers designed a new type of furnace that allows more efficient combustion of coal and better air circulation. These advanced furnaces, often used to generate electricity, have special fire grates called fluidized beds. The grate—or fire bed—is composed of a solid piece of sheet metal. Air pipes perforated with holes are attached to the top of the sheet and blanketed with sand or finely ground limestone. Coal is laid on the top and set afire. Air blown through the pipes escapes through the holes and travels upward through the sand or limestone. The air supports the burning process. The sand, constantly shifted by the force of the air currents, distributes the heat evenly over the fire bed. If limestone is used under the burning coal (in a fluidized limestone furnace), some of the sulfur released by combustion is absorbed immediately by the limestone ($CaCO_2$) to make nonpolluting calcium sulfate ($CaSO_4$). However, a covering of sand (as used in a fluidized sand furnace) might be a better choice for many industries. This type of furnace can use a wide range of fuels. Coal, solid waste, or trash—even wet garbage—can be burned in such furnaces.

Smokestack technology has also improved. Both soot and sulfur dioxide are reduced when smoke is passed through a scrubber. Several forms of smoke scrubbers are in use. In one version, a set of metal screens directs the smoke to an adjacent horizontal chamber. Here, the smoke is sprayed with water or with a water solution of neutralizing chemicals. In other systems, the smoke is directed into the bottom of a nearby tank and bubbled up through a cleansing solution. As the smoke passes through the solution, many pollutants are collected in the liquid. When it can absorb no more pollutants, the solution is drained into a pool, and the solids settle to the bottom. The water—which contains the dissolved chemicals—then flows through an electrically charged screen. The electric charge attracts the molecules of carbon compounds and sulfur dioxide that resulted from the combustion of fossil fuels. The now-purified liquid is recycled. Chemicals collected from the pools and the screens are sold to other industries.

These new techniques can transform former industrial waste into valuable products. Sulfur dioxide gas can be collected and transformed immediately into sulfuric acid. This acid is used in industrial processes such as the manufacture of dyes and paints. Where oil is used as a fuel, ammonia is retrieved by stack scrubbers and provides the raw material to produce fertilizers.

Engineering technology is not the only advanced method of pollution control. Researchers in Europe and elsewhere are focusing on other approaches to the problem of acid rain. For example, Scandinavian scientists are looking for new ways to control the effects of acid rain after it has fallen. In these countries, damage results when winds carry pollutants from the highly industrialized areas to the east and south of Scandinavia. The acid rain falls into Scandinavian lakes and destroys the aquatic plants that feed the fish. Scientists are attempting to reduce the acid concentration in lake water by adding pulverized limestone. They believe that the limestone will neutralize the acids and make the water more hospitable to life.

7

Ozone Depletion

Ozone is a strange substance. Depending on the altitude, it can be life-giving or deadly. When ozone is present at or near the surface of the earth, it is toxic to plants and animals. At that level, ozone is corrosive and acts much like an acid. It causes eye irritation, raw throats, and lung inflammation. For those with asthma or other respiratory diseases, ozone can be fatal. High concentrations of ozone are found in photochemical smog. This type of urban pollution results from a mix of automobile exhaust fumes, bright sunlight, and stagnant air. In 1996, research workers from the Harvard School of Public Health estimated that photochemical smog in the United States is responsible for more than 50,000 hospital cases a year.

In sharp contrast, ozone in the stratosphere—the atmosphere above 30,000 feet (9,000 meters)—is vital to the well-being of every plant and animal on the surface of the earth. The ozone in the upper atmosphere shields the planet from invisible but harmful ultraviolet rays of the sun. If such rays hit the earth full force, they would kill fish and shrimp larvae near the surface of the oceans, stunt the growth of plants, and contribute to vision problems and skin cancer in humans.

The chemical nature of ozone was discovered in 1840 by a German chemist, Christian Schonbein. He found that ozone is a molecule composed of three oxygen atoms (O_3). At the time, electrical discharges were thought to transform ordinary oxygen (O_2) into ozone. Both natural discharges such as lightning or artificial discharges such as sparks made by electrical laboratory equipment were regarded as possible causes. Schonbein found that an ozone molecule was rather fragile because it soon reverted spontaneously to a molecule of ordinary oxygen.

In 1881, a British chemistry teacher, W. N. Hartley, very carefully experimented with ozone. He found that ozone in the atmosphere absorbs a certain portion of the sun's radiation. Charles Fabry, a French physicist, confirmed Hartley's findings in 1913. With the use of high-altitude balloons, Fabry established that ozone was concentrated in the stratosphere at a height of about 60,000 feet (18,000 meters). In subsequent research, chemists and physiologists found that the filtering action of ozone was beneficial to life on earth. They concluded that advanced life-forms could not have developed if the harmful portion of the sun's radiation had not been blocked. In other words, the ozone shield is essential to life on this planet.

Early Warning

In the late 1960s, the political movement to protect the environment was beginning to gather strength. Many people were afraid that their home planet was being spoiled. The overuse of insecticides such as DDT was debated in Congressional hearings and in the law courts. Public concern was aroused by the growing problems of urban smog and acid rain.

Scientific advisers to the U.S. government encouraged the National Science Foundation (NSF) to sponsor a broad international survey on the condition of the atmosphere. Officials of the NSF recruited Carroll L. Wilson and W. H. Matthews. The two men were prominent members of the faculty at the Massachusetts Institute of Technology. In turn, Wilson and Matthews gathered a team of scientists to report on the status of environmental research and the contents of scientific literature on the subject. The Ford Foundation, the American Conservation Association, and the Alfred P. Sloan Foundation provided additional funding for this project. A report, published in early 1971, covered a wide range of environmental problems ranging from photochemical smog to the disposal of solid wastes such as household garbage.

Later in 1971, staff members of the United Nations (UN) held an international conference in Oslo, Norway, to review the state of the worldwide environment. Organizers included personnel from the World Meteorological Organization and the International Council of Scientific Unions—both affiliated with the United Nations. The reports edited by Matthews and Wilson provided the conference members with material for their discussions.

In late 1971, shortly after the UN meetings, NASA scientists declared that exhaust products from the proposed space shuttle might injure the ozone layer. At the time, various rocket engines were being tested by NASA engineers. They determined that all of the engines generated exhaust products that contained oxides of nitrogen and small amounts of chlorine compounds. For the first time, the stratosphere might be polluted by potentially dangerous, man-made chemicals.

The year before, a Dutch chemist and meteorologist, Paul Crutzen, (now working at the Max-Planck Institute for Chemistry), had established that nitrogen oxides could attack ozone. He also discovered that a single molecule of nitric oxide in the stratosphere could set up a chain reaction and destroy many molecules of ozone. The concerns of the space scientists were given some coverage in newspapers and newsmagazines. However, the general public was not interested. The shuttle program was still far in the future.

Another possible environmental problem involved the flight of supersonic transport aircraft (SST). In the early 1970s, this type of aircraft was in the initial stages of development by manufacturers in the United States, England, France, and the Soviet Union. Advocates imagined a fleet of thousands of these aircraft carrying passengers and freight at high speed on long-distance flights. However, environmentalists saw the negative side of an SST fleet. From the publicity surrounding the space shuttle problems, they learned that exhaust fumes from the SST jet engines would generate large amounts of nitrogen oxides. These were the chemicals that Crutzen had shown to attack and destroy ozone. The SSTs would reach the stratosphere, bringing injurious exhaust products near the ozone shield. This time the public was roused to action. Conservationist groups such as the Sierra Club and the World Wildlife Federation sent protests to the U.S. Congress.

Officials of the National Aeronautics and Space Administration, the Department of Defense, and other government bodies finally decided that the SST was not a sound investment. They were not so much impressed by the possible damage to the ozone layer as they were by revelations that the SST would never make a profit. The Soviets had pursued the development of an SST, but abandoned their efforts after their first aircraft crashed at an international airshow. However, a team of British and French engineers designed the Concorde, an SST that flies beautifully. Several such aircraft were built by a combination of French and British companies. As predicted, however, the Concorde is economically unsound and has lost money for 20 years. Therefore, few SSTs have been built, and their threat to the environment is small. The ozone layer was further protected when NASA engineers modified the composition of the shuttle fuel to minimize the release of oxides of nitrogen.

Without a threat from the space shuttle or the SST, the future seemed safe for the ozone layer. Nevertheless, scientists continued to investigate various aspects of the upper atmosphere. In 1970, James Lovelock, an English scientist, was investigating high-level air currents. While studying large air masses such as the jet stream, he encountered the chemical trichloro-monofluoro-methane. For convenience, this family of chemicals is known as chlorofluorocarbons or CFCs. The molecule consists of one central carbon atom linked to two, or sometimes three, chlorine atoms.

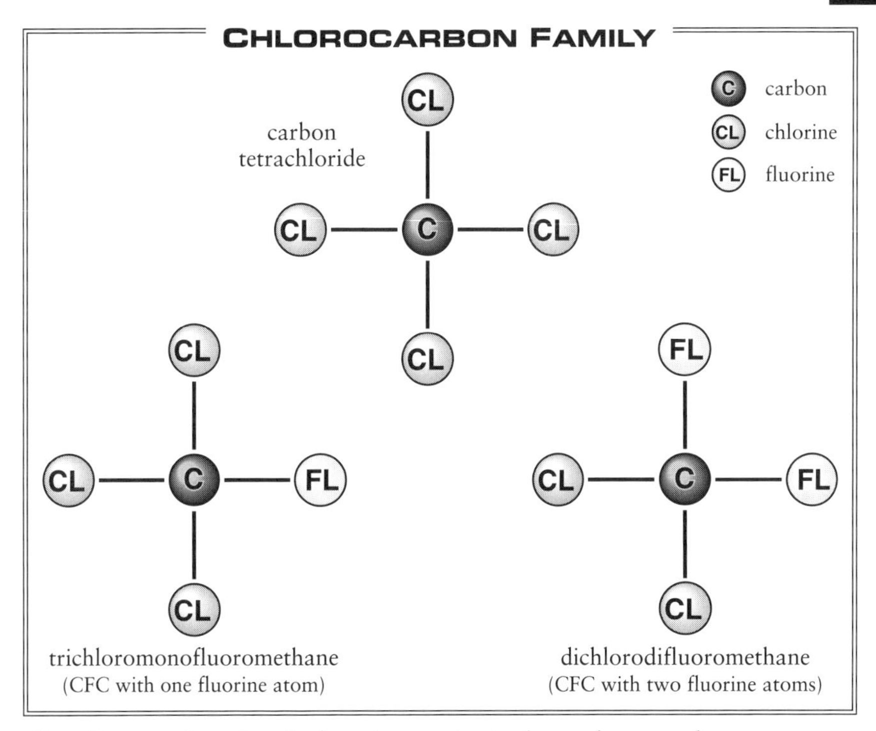

The chlorocarbon family has three principal members: carbon tetrachloride and two chlorofluorocarbon (CFC) molecules, trichloromonofluoromethane and dichlorodifluoromethane.

James Lovelock was not alarmed by the discovery of CFCs in the stratosphere. He had been assured by fellow scientists that this chemical was extremely stable. CFCs had been invented in 1930 by Thomas Midgley, Jr., the same man who developed leaded gasoline. While employed by the General Motors Corporation, the chemist had been searching for a gas to use in automobile air conditioners. The gas needed to be highly stable, nontoxic, and compressible. Midgely knew that carbon tetrachloride was a common, highly stable liquid with four chlorine atoms attached to one carbon atom (CCl_4). He reasoned that by substituting a lighter element, fluorine (F), for some of the heavier chlorine in the carbon tetrachloride molecule, he could make a stable gas. He was correct. Thus, the gaseous CFCs were born.

At first, CFCs were used almost exclusively in air-conditioning and refrigeration systems. Soon, the gases were found useful for cleaning oil or grease from delicate metal parts. Food processors tried CFCs for flash freezing foods. The idea worked well because CFCs were nontoxic and did not react with the foods or any other chemicals.

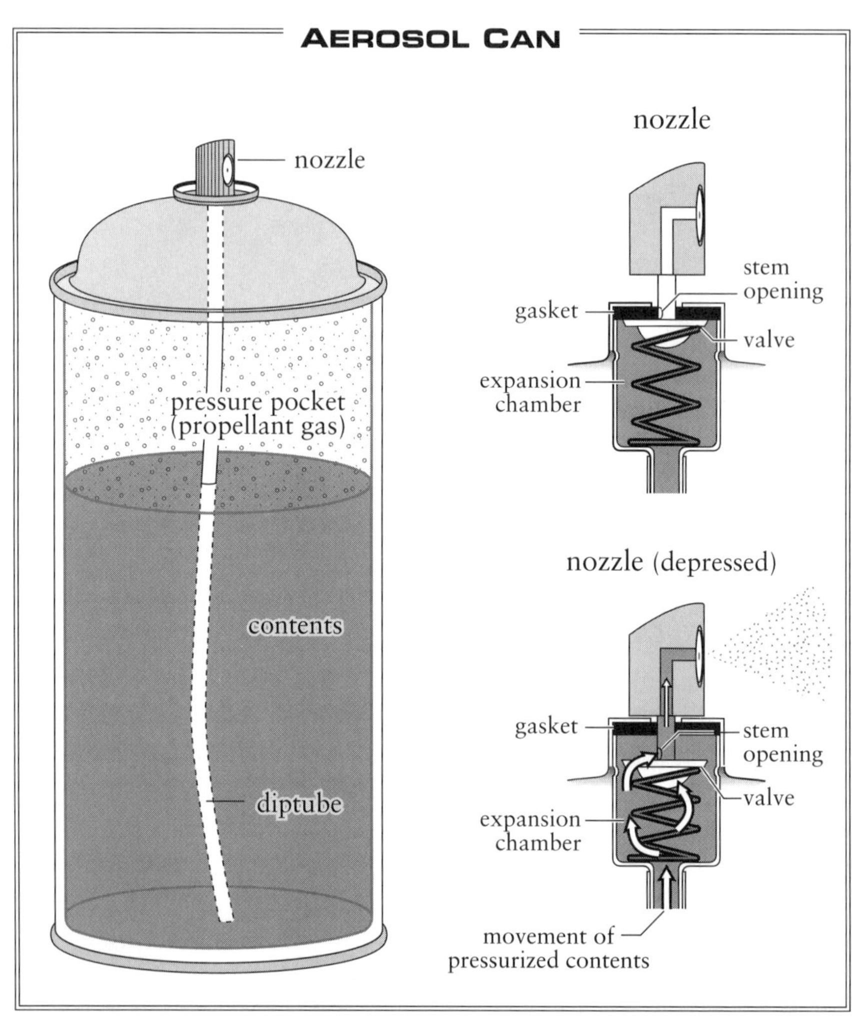

Pressing the button on the top of this aerosol can releases the vaporized contents as a spray. Such cans are still in use, but no longer use CFCs as the propellant gas.

In the late 1940s, Robert H. Ablanalp invented a device called an aerosol dispenser. The dispenser could be installed in a canister containing a propellant gas. At the touch of a button, the propellant forced a portion of any liquid inside the canister to escape through a small nozzle. Depending on the desired consistency, the liquid was dispensed in the form of a strong spray such as hair spray, or as a foam such as shaving cream. The dispenser device could be mass-produced so that each one cost a very small amount of money. The invention became immediately popular. Deodorants, paints, insecticides, whipped cream—indeed, anything that could be sprayed—was soon being dispensed by Ablanalp's device. In 1973, almost 3 billion aerosol containers were sold in the United States. The propellant gases were CFCs, which were ideal for the purpose.

The Ozone Killer

In 1971, F. Sherwood Rowland, a chemist at the University of California at Irvine, became curious about Lovelock's discovery of CFCs in the stratosphere. Because CFCs are relatively heavy gases, Rowland was surprised that they had been found so high in the atmosphere. Although they were known as stable gases, he wondered whether the CFCs in the upper atmosphere would break down under the full impact of the sun's energy. Rowland's area of specialization was the chemistry of irradiated atoms, and he became interested in the long-term stability of CFCs.

Rowland was born in 1928 in Delaware, Ohio. He remained in Delaware to attend Ohio Wesleyan University, where he played varsity baseball and basketball. His graduate work was done at the University of Chicago. There he studied under Willard F. Libby, who won a Nobel Prize for his work on radioactivity.

After receiving his Ph.D., Rowland was awarded a four-year research appointment at Princeton University. He then taught at the University of Kansas. He was only 36 years old when he was

F. Sherwood Rowland discovered that chlorine was released by the CFC molecules that reached the stratosphere. (Courtesy of Felipe Vasquez and the University of California at Irvine)

invited to become the chair of the new department of chemistry at the University of California at Irvine (UCI).

Rowland had established a reputation for being a careful, unbiased research scientist who was interested in environmental issues. Early in his career at UCI, California officials asked him to determine the degree of mercury contamination in tuna caught off their coast. Environmentalists were convinced that the level of industrial contamination was high. They believed that consumers were being exposed to mercuric poisoning by eating the California tuna. After a careful study of the fish, Rowland determined that they contained no more mercury than fish caught in the 1940s.

In 1972, Rowland invited Mario Molina, a Mexican scientist, to join an investigation of CFCs. Molina had recently completed his Ph.D. in atmospheric chemistry at the University of California at Berkeley. The two researchers planned the study together.

Meanwhile, another project was under way at the University of Michigan in Ann Arbor. Research engineers Richard Stolarski and Ralph Cicerone were still intrigued by the possibility that the exhaust from the space shuttles might damage the ozone layer. Because other chemists were already studying the effects of nitrogen oxides on ozone, Stolarski and Cicerone decided to investigate the effects of free chlorine. This choice seemed logical because, at that time, small but detectable amounts of chlorine compounds were found in emissions from solid rocket fuel. The officials at NASA agreed to provide funding for their study.

Stolarski and Cicerone found that free chlorine actively destroys ozone. Furthermore, after a chlorine atom attacks and destroys a molecule of ozone, the same atom of chlorine remains free to destroy other ozone molecules. In short, a small amount of chlorine can destroy a large amount of ozone.

In 1973, the results of this research were presented at a NASA-sponsored meeting in Kyoto, Japan. This report and many others were summarized in a special edition of the *Canadian Journal of Chemistry*. The findings of Stolarski and Cicerone were confirmed by a similar study reported by Michael McElroy and Steven Wofsy, atmospheric chemists at Harvard. The two reports caused little concern because, by this time, chlorine-generating rocket fuels were no longer used. The danger of chlorine in the stratosphere seemed over. Indeed, the prestigious journal *Science* rejected a follow-up report on this danger because the editors thought that it lacked significance.

At about the same time, Rowland and Molina began to obtain troubling results from their investigations on CFCs. They discovered that a CFC molecule did not remain intact—and harmless—when bombarded by undiluted sunlight

Mario Molina was Rowland's partner in the research on CFCs. (Courtesy of Felipe Vasquez and the University of California at Irvine)

in the stratosphere. The CFC molecules broke down and released free chlorine.

Scientists were now ready to tell a disturbing story about a seemingly harmless chemical. When released at ground level, the CFC gas from spray cans and other sources remains intact and does not react with other chemicals. Slowly but surely, these gases swirl upward through the atmosphere. Some reach the stratosphere, where direct sunlight breaks up the CFC molecules. The free chlorine released from the gas then attacks and destroys many ozone molecules. As the ozone layer is depleted, more harmful ultraviolet radiation reaches the surface of the earth.

Breaking the News

After Rowland and Molina submitted their paper for publication, Rowland took a short working holiday in Europe. He met Paul Crutzen in Vienna and disclosed the findings of his research. To ensure the accuracy of his conclusion, Rowland and Crutzen reviewed the mathematical calculations in the study. They both found the same small but important mathematical mistake. The next day, after the error had been corrected, Rowland made his presentation to a group of European colleagues. The scientists were dismayed by the news that the ozone layer was in danger.

Because of their parallel discoveries, Rowland and Crutzen now became allies in the battle to save the ozone layer from destruction. Each had contributed a major piece to the ozone puzzle. In 1970, Crutzen had proven that nitrogen oxides could deplete the ozone shield. Three years later, Rowland had found that the chlorine from CFCs was able to do the same deadly work. After Rowland's research was published in the June 1974 issue of *Nature*, the two men were deeply involved in the struggle to protect the planet.

At first, the general public was indifferent to the article by Rowland and Molina. The report by Stolarski and Cicerone had received the same lack of interest. Indeed, atmospheric scientists seemed to be the only ones concerned by the startling news. Soon, however, attitudes began to change. At a September meeting of the American Chemical Society, Rowland presented a follow-up report on the dangers of CFC. Typically, such meetings are attended by reporters from the major news media. In this case, science reporters from the *New York Times*, *Newsweek*, and *Time* were present. These writers are specialists in making scientific material interesting and understandable to nonscientists. Their articles are read by members of Congress and other high government officials, among others.

Sometimes, a public official responds to one of these issues by making a public comment or a short speech. In the case of Rowland's findings, the response was more vigorous. Representative Les Aspin from Wisconsin immediately submitted a bill to restrict the use of CFCs in aerosol cans. The proposed bill was assigned to a House committee for deliberation, but the session of Congress expired before hearings could be held. This meant that the bill died automatically. Therefore, a new draft of the bill had to be submitted early the following year.

The bill was resubmitted and, in early 1975, the hearings began. Many members questioned the likelihood of ozone depletion, the severity of the threat, and the government's role in such an issue. Following their usual pattern, the members of the congressional committee requested additional studies.

At the urging of Congress, the executive branch of government established the Inter-Agency Task Force to investigate the dangers of CFCs to the ozone shield. The task force, composed of members from 12 government organizations, was led by the representative of the National Aeronautics and Space Administration (NASA).

Measurements of stratospheric ozone are made by instruments on the ground. A device using a special light meter measures the amount of dangerous radiation reaching the surface of the earth. Low readings on the light meter indicate few harmful rays are reaching the earth. High readings show the penetration of more radiation and mean that the ozone shield is being depleted.

To extend and verify the findings of the ground-based apparatus, NASA satellites were sent above the stratosphere. The space vehicle's instruments had been programmed to sample and analyze ozone and other atmospheric components. With few exceptions, the satellite information confirmed the findings of the ground-based instruments.

Congress soon requested the National Academy of Sciences (NAS) to assemble a special panel to review the evidence on ozone depletion. Because the NAS had already established a standing committee on air quality, the task was relatively simple. The staff of the National Research Council, a subdivision of the NAS, quickly organized the panel, and the members set to work.

In June 1975, the report of the Inter-Agency Task Force was presented to Congress. The task force confirmed the findings of Rowland and Molina. That is, their report acknowledged that chlorine was released from the CFCs in the stratosphere. Although the danger was acknowledged, many questions remained unanswered. One question concerned the whereabouts of billions of CFC molecules that had been released over the preceding years. In 1974, scientists had calculated that almost all the molecules were someplace in the atmosphere. No one knew, for example, what percentage of the CFCs had reached the high stratosphere—the level of the ozone shield. Consequently, the members of the task force could not make a final decision on the threat to human health.

Another report was released in early 1976. The review by the National Research Council (NRC) also verified the danger of CFCs. However, the NRC panel members, too, felt unable to call for the regulation of CFCs without more evidence.

Some of the legislators became tired of waiting for decisive action. They determined to employ existing laws to restrict the use of aerosols. At the time of the NRC report, the Toxic Substance Control Act was pending passage by Congress. Legislators attached a provision to the bill that would require special labeling on all canisters that contained CFCs. This act would, in effect, classify CFCs as toxic substances.

Officials of the Food and Drug Administration also utilized existing laws to restrict the use of CFCs in aerosol cans. They wrote a regulation based on the Federal Food, Drug and Cosmetic Act of 1938. The original law stipulated that all containers of toxic ingredients must carry a prominent warning. The new regulation stated clearly that CFCs were toxic and that every product containing the chemical was to be considered dangerous. Hence, the public saw warnings on hair spray, shaving cream, and, indeed, all aerosols using CFCs. In the meantime, the Environmental Protection Agency established restrictions on the import and export of CFC materials.

Industry Responses

In anticipation of a possible boycott by consumers, the manufacturers of personal care and household products lost their enthusiasm for aerosol sprays containing CFCs. Marketing specialists did not want their companies viewed as contributors to atmospheric pollution. Of equal importance, company officials knew that CFCs were not the only gas propellants that could be used in aerosol cans. They began to search for a new propellant.

On the other hand, processors of frozen foods and manufacturers of refrigeration equipment, air conditioners, and plastics were strongly opposed to any restrictions. CFCs were vitally important to their businesses and seemed irreplaceable. At first, however, these industries were not inconvenienced. The initial target was the use of CFCs in aerosol cans. When the EPA considered a ceiling on all CFC production, the industrialists joined ranks with the manufacturers who produced the CFCs. Together, they formed the Alliance for Responsible Chlorofluorocarbon Policy, a public relations and lobbying organization. The members tried to cast doubt on the science that had identified the ozone threat. They maintained that there was no hard evidence of ozone depletion.

The International Response

Everyone concerned with the CFC problem agreed that the issue was international in scope. Groups such as the Organization for Economic Cooperation and Development (OECD), the North Atlantic Treaty Organization (NATO), and several bodies associated with the United Nations added the issue to their agendas. The discussions produced much talk but little action. None of the highly industrialized countries wanted to cut back on CFC production unless all countries followed suit. All the groups looked to the United States for leadership. The situation became very uncomfortable. Powerful U.S. manufacturers produced most of the world's supply of CFCs and were unwilling to cut back or stop production. Leaders from developing countries where CFCs were produced or used in industrial processes saw the restrictions as an attempt to hamper their industrial growth.

In 1978, the United States finally passed firm regulations to ban the use of CFCs in aerosol dispensers. Sweden, Norway, and Canada passed similar laws. Political leaders of the European Economic Community (now the European Union), set goals for the reduction of CFC production. However, the target dates were far in the future and penalties for noncompliance were light.

An Uneasy Peace

In the mid-1970s, before the ban was in effect, fully 50% of CFCs were used in aerosol cans. To help the environment, many people had stopped using aerosols well before the ban went into effect. After the ban, the majority of people in the United States thought that the problem was solved. They were wrong. One of the problems arose because CFCs do not deteriorate under normal conditions. That means that every ounce of CFC remains in the environment. When an old air conditioner is removed to the junkyard, the metal parts deteriorate and the CFCs are released into the atmosphere. The same process occurs with empty aerosol cans, discarded refrigerators, and, indeed, every other product using CFCs.

Even though the demand for CFCs began to drop before the regulations were enacted, the market soon recovered. Almost all cars and homes were being equipped with air conditioning and CFCs were still valuable for cleaning delicate electronic parts. The world market, too, had slowed temporarily, but soon after the aerosol ban it began to grow again.

The industrialists who lobbied against the ban continued to belittle the evidence about ozone depletion. There *was* some uncertainty in the findings of the atmospheric chemists. The amount of CFC molecules in the stratosphere was difficult to establish. The powerful lobbyists claimed that the breakup and reformation of ozone is the result of natural processes.

Indeed, there is a rhythm to the life cycle of ozone molecules. Ozone is a fragile molecule and reverts quickly to pure oxygen. Since the sun is essential to the formation of ozone, the level drops at night and in the winter. Because the greatest intensity of sunlight is found at the equator, the highest density of ozone is found in the atmosphere above the tropics and the lowest found above the poles.

After the ban on aerosols was in effect, the general public became more complacent about the CFCs. The possibility of questionable scientific results also succeeded in calming their fears. Consequently, in 1980, when the Environmental Protection Agency officials proposed a ceiling on CFC production, the public response was unenthusiastic. U.S. Secretary of the Interior Donald Hodel joked about the issue. He was quoted as saying that people should wear sunglasses and sunscreen to guard against ultraviolet radiation rather than worry about the ozone layer.

Public opinion about ozone depletion began to shift again in 1983. One source of concern was a public scandal about the mismanagement found at the Environmental Protection Agency (EPA). The public had reason to believe that some information put out by the agency was not precisely correct and was biased in favor of large industries. EPA officials had failed to impose regulations on CFCs that had been called for by Congress. This failure had prompted a lawsuit by the National Resources Defense Council.

To regain credibility, the EPA began to work through the United Nations to arrange an international agreement to restrict CFC production. If the United States stood alone in regulating the chemical, other countries would continue production of CFCs. The worldwide threat would persist. To resolve the crises, the United Nations needed to draft an acceptable international agreement.

In 1984, members of the United Nations Environmental Program sought to restrict the production and release of CFCs. EPA delegates quickly supported this undertaking. However, west European allies of the United States were unenthusiastic about production ceilings on CFCs. Although the Americans did not want to anger the Europeans, they upheld the new U.S. regulations. A compromise agreement was prepared.

In the spring of 1985, meetings were held in Vienna, Austria, to ratify the UN agreement. Forty-three nations agreed to sign the new pact. Before the negotiations were completed, however, the arrangement to restrict CFC production was withdrawn. Delegates from western Europe and Japan had resisted the provision and the document was ineffective. The members pledged only to continue scientific research, share information, and reconvene in the future.

The outcome was frustrating to such advocates of reform as Sherwood Rowland. Rowland had often appeared in public forums to defend tight restrictions on CFC production. He and other environmentalists despaired that world production would not be regulated until the depletion of the ozone layer was obvious to everyone. By then, it would be too late to save the planet.

A Crisis of the Right Size

As early as 1982, advanced warnings of a new crisis were recorded by the British Antarctic Survey team. The team collected daily readings from special light meters that measured the sun's dangerous radiation. During September and October—spring months in Antarctica—the meters registered large increases in the amount of radiant energy that reached the surface of the earth. The team members feared that the ozone layer had thinned appreciably. At first, the team coordinator, Cambridge University scientist Joseph Farman, was highly skeptical about the accuracy of the information. Farman worried that the equipment was giving false measurements. He did not want to publicize the findings until the team was confident of its conclusion. If the scientists could not prove their contention, the negative publicity might put the survey out of business. The sponsor, the British government, would be disturbed if the scientists published an unverified account of a possible ozone crisis. At the time, British industrial interests were fighting against restrictions on CFC production.

Farman decided that he must confirm the findings before he released a report. In 1984, he compared his readings with those from an installation on the other side of the Antarctic continent. The results were almost identical. Although the report might anger the sponsors, Farman knew that he must publish his findings. On Christmas Eve 1984, he sent the document to *Nature*, the important British science journal. The article was published in May 1985.

Scientists in the United States were astonished by the news. Few had heard of Joseph Farman or the British Antarctic Survey. Cicerone and other atmospheric chemists wondered why the ozone reduction had not been detected by the NASA satellites and recorded by NASA computers.

In fact, corroborating information had been collected and reported. A researcher at the Japanese Antarctic Research Station had observed a month-long record of ozone depletion. In

1984, the Japanese scientist reported his findings at a small research meeting in Greece. The significance of this information was not recognized by those at the meeting. Ironically, the members included Sherwood Rowland.

Both humans and computers have an imperfect ability to analyze information. Scientists soon discovered NASA's computer problem. The computer had been programmed to analyze information sent from the satellites. However, to avoid false readings, the computer was programmed to ignore measurements that greatly deviated from the average. Although the low ozone count over Antarctica had been properly recorded by the instruments, the unusual measurements had been disregarded by the computer and had not appeared on the printouts. When the satellite readings were re-evaluated, scientists learned that a hole in the ozone layer had first appeared in 1979. By 1984, the hole had grown to be the size of the United States. The level of ozone within the hole was even lower than the atmospheric chemists had predicted. They had anticipated a 15% decrease in ozone but the measurements showed fully 50%—an awesome drop.

No Retreat

In the summer of 1986, news of the ozone hole was reported by the media. Even with this evidence, the CFC manufacturers did not admit defeat. Their publicists stated that the ozone hole was a temporary abnormality. They were convinced that the next set of readings—due in September and October—would prove that the ozone layer was intact.

Shortly after the British and Japanese findings were made public, the National Oceanographic and Atmospheric Administration (NOAA) organized an expedition to the Antarctic. The group arrived during August—late winter in the Southern Hemisphere.

They recorded ozone data during the South Pole's spring months of September and October. Unfortunately for the manufacturers, the expedition gathered information that proved the reappearance of the hole. Fortunately for the environmentalists, news of the Antarctic hole captured the attention of the public. NASA published impressive diagrams based on measurements recorded by the satellites. The images made it clear that something very strange was happening to the ozone layer. The public wanted immediate action.

The manufacturers continued to claim that chlorine from CFCs did not cause the hole. They maintained that more research was needed to resolve the issue. Indeed, atmospheric chemists such as Ralph Cicerone were also cautious. Cicerone wanted another year of observations before making a final decision about the ozone-CFC problem.

This bleak image of the U.S. base at McMurdo was taken in the summer in Antarctica. It is not hard to imagine the desolation at other seasons of the year. (Courtesy of the U.S. Geological Survey, Special Committee on Antarctic Research)

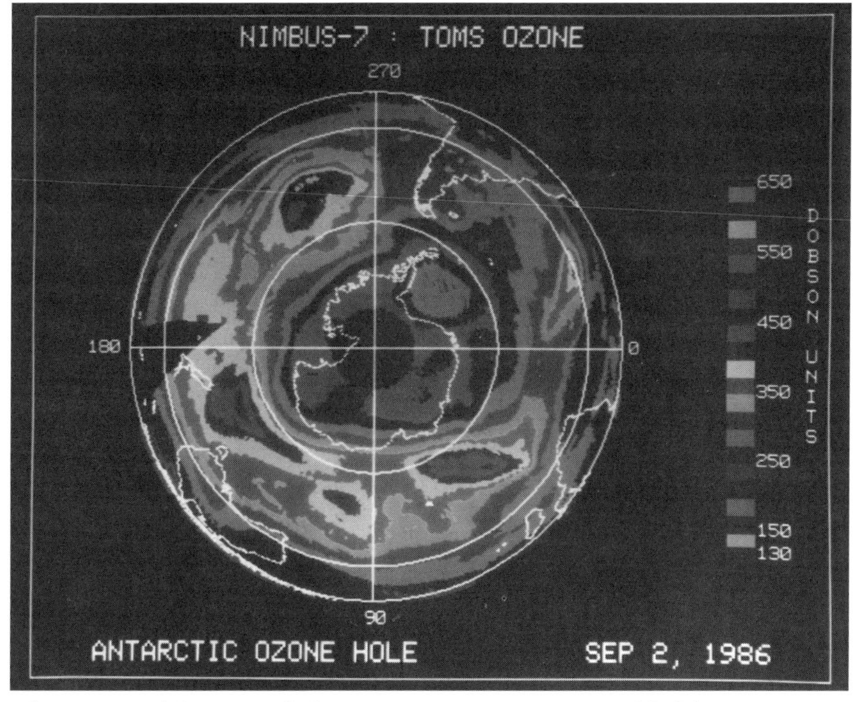

This image of the ozone hole over Antarctica was assembled from measurements made by satellites. (Courtesy of the National Aeronautics and Space Administration)

The United Nations Comes to the Fore

Even before reports of the Antarctic ozone hole in the summer of 1986, the UN Environmental Program and the U.S. Environmental Protection Agency were holding a series of meetings. These technical and political discussion groups meet in Leesburg, Virginia. They drafted severe regulations to limit CFC production. With this in mind, manufacturers conceded that some regulations were necessary. Undoubtedly, this decision reflected their self-interest. The industrialists knew that a substitute for CFC would eventually be found. Each hoped to discover and market the replacement.

Another important conference began in September 1987. The UN Environmental Program was scheduled to convene in Montreal, Canada. Some U.S. delegates were still undecided. However, George Bush, then U.S. vice president, advocated an international agreement to suppress the production of CFCs. The United States was now a leader in the movement. Even so, the conference soon bogged down. To avoid the embarrassment of a totally unproductive meeting, last-minute compromises were made. After the revisions were completed, the first international environmental agreement was signed by 57 countries in September 1987. Skeptics remarked that the most productive outcome about the agreement was the decision to continue research on ozone depletion.

The Big Push in the Antarctic

While the Montreal meetings were under way, a second and larger Antarctic expedition was being organized by NOAA with technical collaboration from NASA. The scientists were given the support of light meters on the ground, satellite observations overhead, and a very high-altitude NASA research plane, the ER-2. The research plane was scheduled to fly into the ozone hole at an altitude of 60,000 feet (18,000 meters). The flight was dangerous because the ER-2 has a single engine and had never been flown near the South Pole, where temperatures drop to -130 degrees Fahrenheit (–90 degrees Celsius). Although the instruments did not work properly on the first flight, the second flight was a total success.

By using information gathered from the ground, the satellite, and the plane, the team successfully measured the chlorine molecules in the Antarctic ozone hole. All of the readings were high. In fact, the concentration of chlorine was 300 times greater than expected. The connection between chlorine and the ozone hole was firmly established.

The ER-2 was flown near the South Pole in extremely low temperatures to determine how much free chlorine was present at stratospheric levels. (Courtesy of the National Aeronautics and Space Administration)

Ironically, the explanation for the Antarctic hole had been proposed by Ralph Cicerone shortly before this discovery. He had mathematically demonstrated that chlorine molecules in the stratosphere could attach themselves to microscopic ice crystals. When the Antarctic spring sun reached these crystals, the ice would melt and the chlorine would be released. It could promptly attack the ozone molecules. The only possible source of chlorine was the CFCs.

Final Victory

The publicity about the Antarctic ozone hole solidified public opinion in favor of restricting CFC production. Most people thought that CFC manufacturers must give up their battle against regulations. Undaunted, the industrialists continued to deny that CFCs were depleting the ozone layer. However, their own industrial chemists recognized that the struggle was useless. The public would become hostile if the companies did not accept the facts. In mid-March, 1988, the main U.S. producer of CFCs announced that production would soon cease.

More than seven years later, on October 11, 1995, Sherwood Rowland, his partner, Mario Molina, and their European collaborator, Paul Crutzen, were awarded the Nobel Prize for chemistry in a ceremony in Stockholm, Sweden. For 20 years, Rowland had spoken to indifferent or skeptical audiences about the ozone crises. The Nobel Prize was a fitting recognition for his diligence.

8

Global Warming

Many scientists believe that the earth is getting warmer. The warming trend is thought to be the result of a condition called the greenhouse effect. Although the ozone layer screens out most of the Sun's harmful ultraviolet rays, atmospheric gases allow the Sun's beneficial radiant energy to reach the surface of the Earth. When the sun's radiant energy reaches the surface, it acts to warm whatever it touches. As soon as the warming begins, the heat energy starts radiating back into the cooler atmosphere. In an apparent contradiction, the same gases that allow the sun's rays to enter the atmosphere also block much of the newly generated heat from ascending into outer space. Indeed, the gases in the air reflect the heat back toward the ground. This allows the planet to retain warmth and support life. The gases act in much the same manner as the glass walls and roof of a gardener's greenhouse. They allow sunlight in, but help keep the heat from escaping. Thus, this process is called the greenhouse effect.

This effect is noticeable on warm late spring or early summer days. After a clear, bright, sunny day, temperatures in the Middle Atlantic states can easily reach 80 degrees Fahrenheit (27 degrees Celsius). If the skies remain clear at night, the temperature will fall into the 60s (16 degrees Celsius) before dawn. The clear night sky allows much of the surface heat to radiate into space. However, if clouds roll in during the evening, the night will remain warm. Temperatures will fall only into the 70s (21 degrees Celsius). The clouds serve as a blanket—or the roof of a greenhouse. They block the heat radiating from the ground and reflect it back toward the surface of the earth.

Water as a Greenhouse Gas

Water vapor and carbon dioxide gas are chiefly responsible for the greenhouse effect. Whether in the form of clouds or as an invisible gas, water vapor accounts for about 75% of this condition. Water vapor, along with nitrogen and oxygen, is one of the three most abundant atmospheric gases. Therefore, it is not too surprising that water vapor is the most effective greenhouse gas. However, carbon dioxide (CO_2), which accounts for only .0003 of the earth's atmosphere, is the second most important greenhouse gas.

Water vapor and carbon dioxide contribute to the greenhouse effect in different ways. The far greater quantity of water vapor molecules is more effective at blocking the heat from escaping into outer space. However, the carbon dioxide molecules are more efficient at reflecting the heat back toward the surface of the earth so that the much smaller amount of CO_2 makes a significant contribution. Together, the two gases account for more than 95% of the greenhouse effect.

Recently, scientists have proven that the level of atmospheric carbon dioxide is on the rise. Therefore, the greater number of carbon dioxide molecules in the blanket of atmospheric gases will increase both the heat-blocking and heat-reflective ability of the gas. Less heat will be able to escape into deep space.

The anticipated increase of CO_2 will aggravate the greenhouse problem. A warming trend will lead to a more rapid evaporation of water from the world's oceans. This will send more water vapor into the atmosphere. The presence of more water vapor will improve the heat retention of the atmospheric blanket. The greenhouse effect will increase, and the warming trend will continue at a faster pace.

Carbon Dioxide

In the spring of 1755, a Scottish scientist named Joseph Black identified carbon dioxide as a gas with a specific molecular construction. More than 100 years later, in 1896, Svante Arrhenius, a Swedish physical chemist, performed research that led to important discoveries about CO_2. Arrhenius had spent many years investigating electrical conductivity and found that CO_2 is a good insulator. Toward the end of his career, his interests expanded, and he sought to explain the ice ages. Arrhenius's new studies showed that carbon dioxide acts as an insulator to help maintain the earth's climate. In other words, CO_2 helps retain the sun's warmth. Arrhenius concluded that an increased amount of CO_2 had contributed to the warmer temperatures that melted the glaciers and caused the end of the Ice Age. The idea that carbon dioxide could influence the earth's climate was not easily accepted by his fellow scientists. Subsequent research, however, verified his theory. During the last 15 years of his life, he shared his knowledge and love of chemistry by producing science books for children and nonscientists. Before his death in 1927, Arrhenius was honored with the Nobel Prize for chemistry and other important awards.

Too Much Carbon Dioxide

In 1938, a British mining engineer, George Callendar, completed a 10-year study of the earth's climate. After a careful analysis of his temperature readings, Callendar concluded that the information revealed a general warming trend. The British engineer knew of Arrhenius's report and was convinced this trend indicated an increase of atmospheric CO_2. Most scientists were skeptical of his conclusion. They saw no hard evidence of an increase in carbon dioxide.

By 1958, the warming trend and the CO_2 problem had become more apparent. Scientists from 70 countries participated in a series of meetings called the International Geophysical Year—a project that lasted for 18 months. The participants hoped to establish a set of basic measurements on global climate factors such as temperature, rainfall, and atmospheric carbon dioxide. All future measurements of climate factors would be compared to these basic measurements, and any significant changes would be easy to detect. Roger Revelle, the head of the Scripps Institute of Oceanography at La Jolla, California, led the planning groups.

Revelle and a colleague, Hans Suess, were especially concerned about the possible increase in atmospheric carbon dioxide. They began to investigate whether the oceans would be able to absorb a greater amount of this gas. The scientists hoped that increased atmospheric carbon dioxide would be used by sea creatures to manufacture their shells. However, the scientists found that the oceans and the sea creatures were incapable of removing much additional carbon dioxide. Consequently, Revelle and Suess stressed the necessity of establishing an accurate baseline to monitor the increasing amounts of CO_2 in the atmosphere.

This task was assigned to Charles Keeling. Keeling had designed a delicate instrument that accurately measures CO_2. The sensitive device detects a change of one molecule of carbon dioxide per million molecules of air. In order to establish the baseline measurements of CO_2, Keeling installed one of his instruments on the slopes of Mauna Loa in Hawaii. The site is far from any industrial activity, and the daily measurements were free of contamination.

The CO_2 readings from Mauna Loa have become classic scientific data. The records show a saw-toothed progression that began in 1960 with 320 parts of CO_2 per million and increased

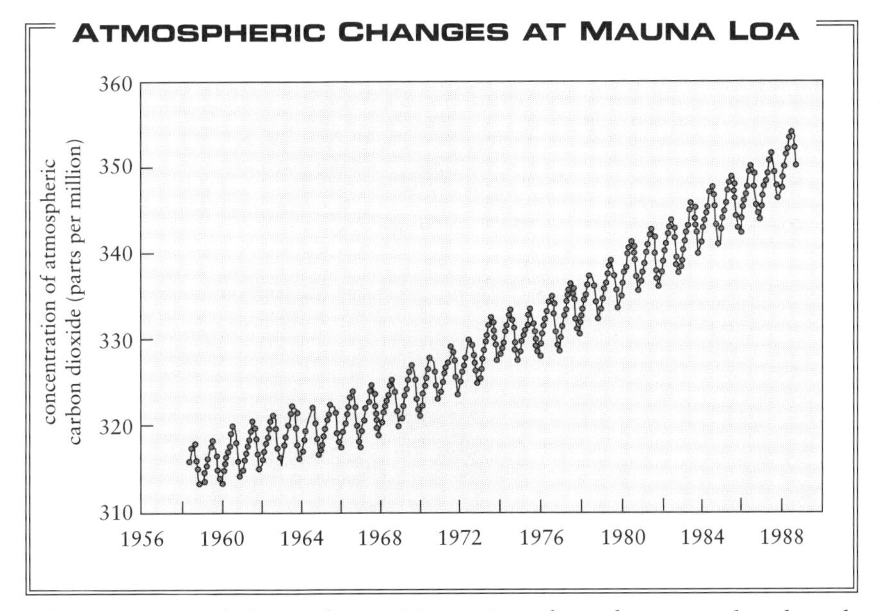

Observations made by Keeling at Mauna Loa show the seasonal cycle and the step-wise increase in the amount of carbon dioxide in the atmosphere.

to more than 360 parts per million by 1995. The Keeling Curve is undisputed evidence that the amount of CO_2 is growing and that the rate of growth increases every year.

General Effects

Scientists agree that global warming will not be uniform around the earth. The greatest warming will occur near the North and South Poles and the least at the equator. The areas between the poles and the equator will show a gradual and intermediate rise in temperature.

Specialists in many countries are studying this greenhouse effect to determine how fast temperatures will rise and how this change will affect life on the earth. Whether the temperature rise is rapid or gradual, millions of people will be affected.

The economies of many industrialized countries depend on electric power manufactured by steam turbines. Coal, oil, and natural gas—the fossil fuels—produce most of the steam. Unfortunately, all fossil fuels create carbon dioxide as one of their waste products. Coal is the worst offender. To make each kilowatt of electricity, coal produces 10% more carbon dioxide than oil and twice as much as natural gas.

If the burning of coal is restricted, the cost of electricity will rise because coal is cheaper than oil or natural gas. If the price of electricity increases, luxuries such as air conditioning may be restricted to the rich. If the electric companies try to maintain existing prices by cutting back on production, power failures will be more frequent. Brownouts, caused by temporarily reducing the amount of electricity, will become commonplace. In short, the lifestyle of countless individuals will become less pleasant because of restrictions on fossil fuels. This possibility is unacceptable to most citizens. Scientists, engineers, and environmentalists are seeking alternative methods to generate electric power without increasing atmospheric carbon dioxide.

Specific Effects

If global warming happens quickly, a serious consequence could be the breakup and partial melting of the polar ice caps. The melt water would cause ocean levels to rise and flood low-lying coastal areas. Unfortunately, many of the world's richest and most populous cities are located near bodies of water. In industrialized countries, the main problem will be property loss. New York City in the United States and Amsterdam in the Netherlands are prominent examples of wealthy coastal cities. In the less industrialized countries, flooding will cause many casualties, destruction of buildings, reduction in agricultural land, and consequent food shortages. Poor cities such as Chittagong in Bangladesh, located on the Bay of Bengal, will experience major problems.

If the sea level rises slowly, human adaptability will be able to minimize the loss of life. However, the loss of property will be extensive. Unless enormous dikes are built, such as those in Holland, valuable buildings and agricultural land will become unusable.

Other Effects

The U.S. Department of Energy has funded studies to determine the effect of a warming trend on human health, agriculture, forestry, and fishing. In the matter of human health, diseases prevalent in tropical or semitropical regions could become commonplace in northern areas. For example, malaria might someday be a health problem in countries that are now too cold to support the tropical *Anopheles* mosquito responsible for transmitting the illness.

If winters in Canada and northern United States become milder, some illnesses such as colds and flu could be alleviated. Of greater importance is the fact that fewer people die of heat-related problems than exposure to extreme cold. Therefore, the warming trend would benefit the poorest and most vulnerable segments of society.

In agriculture, a warmer climate and an increase in carbon dioxide might seem to be a benefit for humankind. A carbon-dioxide–enriched environment causes many plants to grow more rapidly and to a larger size. Growing seasons would lengthen. Land that produces a single yearly crop might be able to grow two successive crops. However, two possibilities cloud this happy prospect. First, soil conditions change from one region to another. Both soil and climate are responsible for the highly productive corn-growing regions of Indiana, Illinois, and Iowa. A warming trend might shift the best growing conditions into northern states such as Michigan. The soil in those states, however, is not as well suited for the cultivation of corn. The yield of this important crop would suffer.

Adverse effects would result if a climate change caused a difference in the pattern of rainfall. Less rainfall could transform productive growing areas into deserts. If the water flowing in the Colorado River and its tributaries decreased in volume, the consequences could be disastrous. Millions of people depend on this river to provide crop irrigation, drinking water, and recreation. Parts of the Southwest, such as southern California, would suffer greatly from any change in the fragile river system.

Global warming would affect forests. Each species of tree grows best in a specific soil and temperature range. Although tree populations would be less affected than farm crops by changing weather conditions, some tree species might become extinct.

The fishing industry would experience extensive changes from a warming climate. In major saltwater fishing areas, the food chain is supported by tiny, aquatic, one-celled plants and animals called plankton. Plankton thrive in relatively cold, nutrient-rich water that has risen to the surface from the depths of the ocean. If the ocean waters become warmer, the numbers and vigor of the small creatures could be reduced. Smaller fish would have less to eat. Large, edible fish would decline in numbers because their food supply would diminish. The fishing industry fears that this possibility might cause the extinction of many popular favorites, and the amount of fish in the human diet would decrease.

El Niño

A study of natural climate and weather variations such as El Niño gives scientists some understanding of man-made environmental problems such as the greenhouse effect. El Niño is a strange meteorological condition that recurs every two to seven years and affects the climate of the entire world. It usually develops off the coast of Peru in the month of November and reaches its peak in December near Christmastime.

Indeed, El Niño is the Spanish expression for the Christ child. El Niño has been noted by Peruvian farmers and fishermen for many decades because its arrival signals a dramatic change in the fish population and an equally dramatic increase in rainfall. Curiously, this peculiar weather condition stimulated little scientific interest or investigation until fairly recently.

Normally, strong trade winds blow the warm surface water away from the coast of central South America and across the Pacific Ocean toward Asia. The enormous volume of water that is swept westward actually causes the ocean level off Asia to measure two feet higher than that off central South America. Near the Peruvian coast, cold, nutrient-rich water rises up from the deep ocean to replace the water swept westward by the winds.

When El Niño comes, these strong, westerly winds weaken, and the warm surface water flows back toward the coast of central South America. The water levels of the eastern and western coasts of the Pacific are equalized. Along the coast of Peru, the normal upsurge of cold water is inhibited by the presence of warm surface water from the mid-Pacific. Peruvian fishing is immediately affected because the surface water does not carry nutrients to the coastal fishing grounds.

In the early 1920s, Sir Gilbert Walker, a British scientist, began investigating weather conditions in the western Pacific between Australia and Indonesia. At first, Walker's work indicated that atmospheric pressure in the region was uncommonly stable. He found that high atmospheric pressure is usually recorded along the southwestern rim of the Pacific and low pressure along the southeastern rim. Soon, however, Walker realized that every few years, the pressure readings are reversed during the months of November and December. During those months, low atmospheric pressure is recorded in the southwestern Pacific and high pressure in the southeast. Walker called the seesaw phenomenon the Southern Oscillation. No one associated these meteorological findings with the strange rhythmic occurrence of El Niño. It was not until 1966 that Jacob Bjerknes perceived the connection.

Jacob Bjerknes was a brilliant weather forecaster from Norway who founded the Department of Atmospheric Science at the University of California at Los Angeles. He was the first person to understand the workings of the El Niño effect. (Courtesy of Eugene Rasmusson, University of Maryland)

Jacob Bjerknes and his father, Vilhelm, are famous for the mathematical and graphical representations used in weather forecasting. In the early 1940s, Jacob Bjerknes traveled to the United States from Scandinavia to give a series of lectures.

During that time, the German army invaded and quickly conquered his native country of Norway. Bjerknes was trapped in the United States. He was invited to join the faculty at the University of California at Los Angeles and soon founded the Department of Atmospheric Science there.

In the early 1960s, Bjerknes was engaged in a long-term study of the relationship between the oceans and the atmosphere. While analyzing conditions off the coast of Peru, he noted a recurring event. In most years, high barometric pressure is prevalent during November and December. During these years, strong prevailing winds blow from east to west. However, every few years, the barometric pressure during those same months is dramatically reduced, and the winds blow from west to east. Local fishing and farming are disrupted, and strange weather conditions are reported.

Bjerknes was familiar with Sir Gilbert Walker's research on atmosphere pressure in the western Pacific. He realized that his observations along the eastern coast of the ocean echoed the cyclic pressure changes recorded by Walker. El Niño was recognized as a worldwide phenomenon.

The impact of El Niño is large and varied. Scientists believe that an investigation of these changes will reveal possible consequences of future global warming. Strategies that modify the adverse results of El Niño can be used to modify the greenhouse effect. For example, rainfall in the usually semiarid zone near the west coast of Peru increases dramatically during El Niño years. Improved forecasting now informs farmers of the upcoming weather. To benefit from El Niño's rain, they can cultivate a moisture-loving plant such as rice that season. Similar variations in agricultural practices will be necessary to survive a global climate change.

The fishing industry must also expect disruptions during El Niño. Catches of cold-water fish, such as sardines and anchovies, decline. However, other popular seafoods, like scallops, are more abundant. In 1997, the sea off California experienced a profusion of nonnative, warm-water fish. Those who engage in sports fishing were overjoyed by the prospect of catching large numbers of unusual fish. Presumably, the effects of El Niño were responsible for currents of warm water that brought the fish. Perhaps, displaced fish will be one of the results of a global warming.

Many effects of El Niño are damaging to people and crops. In some areas, excessive rain causes temporary lakes to form in low-lying land. The standing water expands the breeding grounds of the *Anopheles* mosquito and adds to the risk of malaria. The rains also promote the growth of wild plants that provide abundant food for destructive insects. Insects such as grasshoppers soon show an increase in both size and numbers. El Niño is responsible for other unusual weather patterns. Expected rainy seasons are sometimes replaced by drought. This weather condition causes crop failures, reduces the availability of drinking water, and permits forest and brush fires in the western Pacific islands.

Automatic Recovery Processes

Some argue that the effects of global warming will be insignificant and few people will notice the difference. The famous Gaia theory holds that the earth will adapt to change and continue to support life. Gaia, a very early Greek deity, was worshipped as Mother Earth and goddess of fertility. Indeed, those who follow this theory believe that the effects of global warming may benefit the earth.

Green plants use carbon dioxide as raw materials for growth and development. If the carbon dioxide level increases, many plants will grow faster and reach a larger size at maturity. The larger, healthier plants will absorb more CO_2. This absorption of CO_2 will help the environment. At the same time, the improved plants will supply more food for the world population.

According to the Gaia theory, aquatic plants and animals will also benefit from more CO_2. Carbon dioxide will increase the growth of microscopic plants that live in the surface waters of the oceans. These plants will nourish more small creatures that will use more CO_2 to manufacture calcium carbonate, one of the main ingredients in seashells. The larger supply of small plants and animals will nourish more fish for human consumption.

Some argue that the earth will soon enter a cycle of very cold weather. Indeed, the earth has always had a cyclic climate. During frigid cycles, ice sheets—or giant glaciers—spread from the poles and covered major portions of the present temperate regions. Some people believe that the warming effect caused by greenhouse gases will save the world from another ice age.

There are some difficulties with this argument. Natural warming and cooling trends take place gradually over hundreds or thousands of years. However, the effects of increased carbon dioxide are expected to take only decades. Consequently, a rise in CO_2 might hasten the present warming trend. Temperatures may continue to rise for decades or even several hundred years before the beginning of a cooling trend. By that time, the environment would be damaged, and the oceans would have flooded coastal farmlands and cities.

9

Remedies

What can be done to prevent or minimize global warming? Industrialized nations—particularly the United States—must use less energy produced from fossil fuels. Five percent of the world's population lives in the United States. That 5% consumes 25% of the energy produced each year.

The simple act of switching off unnecessary electric lights will save a surprisingly large amount of energy. The use of fluorescent bulbs rather than incandescent bulbs will further increase conservation. Setting air-conditioning systems a few degrees higher in summer and heating systems a few degrees lower in winter will also save large amounts of electricity and fuel. All of these measures will reduce the need for energy. In turn, this reduction will lower the use of fossil fuels that produce atmospheric carbon dioxide.

Public Policy Issues

Both the depletion of the ozone layer and the increase in carbon dioxide are global issues. Indeed, present and future populations will suffer if there is no remedy. No one country can solve the problems. All people in all nations must work together.

Ideally, the United Nations should provide leadership for such a collaboration. However, the UN has a mixed record in dealing with difficult issues. For example, at the beginning of the ozone depletion crisis, the United Nations Environmental Program (UNEP) was consistently slow in responding to scientific findings. UNEP conferences were held yearly, and delegates brought forth a wide range of views. Concluding statements usually called for more research, further information exchange, and additional meetings. Multinational treaties were rare.

In some ways, the situation in the United States parallels that in the United Nations. Citizens have conflicting ideas and interests concerning environmental issues. Moreover, agencies within the federal government advance a variety of ideas on public policy. Politicians and government administrators find it difficult to agree on solutions that require people to change their behavior. Today, federal agencies provide funds to study methods to generate energy by the use of low-polluting or nonpolluting materials.

Technical Responses

METHANE PRODUCTION

One of the most promising techniques is the use of carbon-based waste materials to produce methane (CH_4). Methane, a simple molecule made of one carbon and four hydrogen atoms, is an inflammable gas at room temperature. The gas can be used to fuel power plants and other systems that commonly use fossil fuels.

Natural gas, a cleaner and more efficient fossil fuel than coal or oil, is a source of methane. There are abundant supplies of natural gas in many parts of the world, including areas without coal or oil resources. Methane is easy and inexpensive to produce. Indeed, most carbon-based materials can be converted to methane by the natural action of bacteria.

Backyard gardeners can manufacture methane by mulching. Methane gas is a by-product of this process, in which grass cuttings, leaves, and other household garbage are dumped into large metal containers or pits dug into the ground. Small amounts of fertilizer or other sources of nitrogen are added. Bacteria multiply rapidly in this medium and digest the mixture. After a period of time, the action of the bacteria reduces the material to a moist, porous, brown substance. When added to the garden, the mulch fertilizes and softens the soil.

Some caution is needed when methane is produced because this gas—like CO_2—is also a greenhouse gas. Backyard mulching systems usually allow methane to escape into the atmosphere. Some mulching systems now capture and compress the gas for use in backyard barbecues and winter fireplaces. However, large quantities of unrecoverable methane are produced by the digestive systems of cows and other animals. Unusable methane is also found during plant fermentation in wet rice paddies and other agricultural sources.

Methane is produced commercially by large farms and waste management companies. These organizations use sizable special chambers similar to those found in small backyard systems. Waste materials such as municipal garbage, straw, crushed corn stalks and cobs, and used paper are employed as mulch to

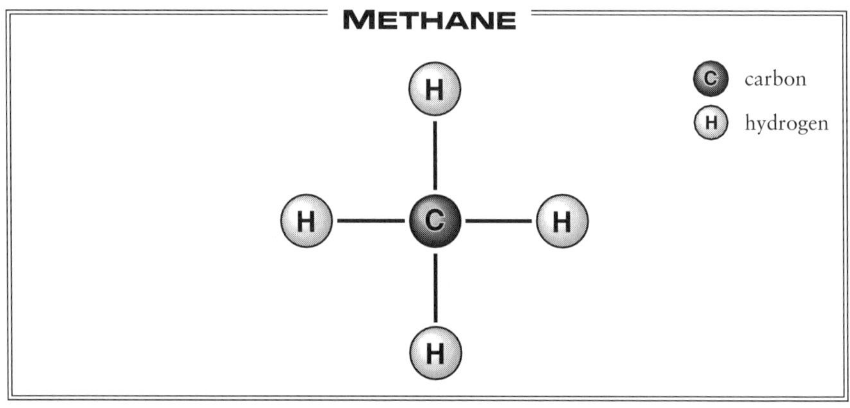

Natural gas is composed of methane, the simplest hydrocarbon molecule.

produce methane. Indeed, this environmentally safe fuel can be quickly and easily produced as long as plants grow on the earth.

Methane and other gaseous fuels have another major advantage. These gases can be used in an extremely efficient and cost-saving technology. Hot gases from the burning fuel cause a turbine to rotate. The turbine is connected to a generator that produces electric power. In turn, the exhaust gases from the same turbine are used to boil water that generates steam. The steam is used to turn another set of turbines and generate more electric power. Thus, gases from the burning fuel drive a turbine that generates energy, and exhaust gases from that turbine produce steam to create more energy. Manufacturers can use the same amount of fuel to produce almost twice the electricity and generate half the pollution.

REFORESTATION

In many parts of the world, programs of reforestation are becoming more prevalent. These programs address a variety of problems such as timber shortage, soil erosion, mud slides, and of increasing importance, the growing excess of CO_2. Trees, and to a lesser degree all plants, require CO_2 to grow and therefore absorb quantities of the gas.

In the United States, reforestation is practiced by timber and paper companies that harvest large numbers of trees. Federal law mandates that these companies replace trees cut on federal land administered by the Forest Service. For public relations, political, and economic reasons, many companies promote reforestation programs. Very large industrial firms often set up reforestation departments within their companies. Others engage small companies that provide the service of replanting forests. These companies specialize in developing tree stocks that are naturally resistant to disease and insect attacks. Their business is booming. One small Canadian company sold almost 400,000 tree seedlings in its sixth year of operation.

In recent years, individual states in the United States have become more active in caring for their timber resources. Several states, notably Washington and Oregon, have regulations restricting logging in sensitive areas. Many lumber companies use a method called clear cutting. In clear cutting, every tree in a designated area—regardless of size or condition—is cut down. The method is popular because the area can then be easily and completely cleared of the felled trees. Although the method saves money for timber companies, it is not good for the environment.

The soil in many northwestern states consists of a thin layer of topsoil that covers loose sand and gravel. Clear cutting damages the root systems that helped retain the thin layer of top soil. Erosion, mud slides, and other ecological problems can result. In the state of Washington, the salmon industry is suffering from the results of clear cutting. Loosened gravel and other matter enter the waterways and silts up, or clogs, streams where salmon spawn. As a result, the salmon population has been dramatically depleted.

Although reforestation offers a solution to some environmental problems, many ecologists have reservations about this approach. Scientists worry that it creates an artificial environment rather than a natural forest. Indeed, reforestation usually creates a monoculture—a forest planted with seedlings of the same age and type of tree. The new trees will succeed in absorbing CO_2. However, the reforested area may not provide the varied habitat needed by the creatures that populate a natural forest.

In the United States, the Sierra Club, the Audubon Society, and other well-known organizations promote tree-planting projects. Smaller groups such as the Global Relief Program are active in urban settings. However, many areas of the world are not interested in such projects. In some countries, more urgent problems give reforestation a low priority. In others, the original deforestation profoundly changed soil and water conditions, and growth of new forests is very difficult if not impossible. Some areas around the Mediterranean Sea fall into this category.

Unused land can revert to a forested area by a natural process. In New England states, a form of spontaneous reforestation can occur. Steep or rocky land reverts to scrub growth when taken out of cultivation. After many years, these plots may become fully diversified forest areas. Although some larger trees may be cut for firewood, the reforested area usually remains in an untamed condition.

In poorer countries of the world, the need for firewood is so great that land can rarely be reforested. Trees are cut down long before they reach maturity. However, in industrialized areas of Europe, uncultivated land is rarely stripped of saplings for firewood. Studies sponsored by the Finnish government suggest that the expansion of northern European forests is responsible for the absorption of 15 million tons of carbon dioxide every year. According to research sponsored by the U.S. Forest Service, spontaneous reforestation in the United States increases the absorption of excess CO_2.

The formation of peat allows forests in the world's temperate zones to retain a goodly amount of the excess CO_2. Peat is made of vegetable matter such as leaves and bark that have absorbed CO_2 during their lifetime. The vegetation falls to the forest floor and is gradually covered by earth. The mixture is slowly compressed by the weight of the soil. Early environmental studies of temperate-zone forests neglected to take into consideration the amount of carbon dioxide held by peat.

Northern, temperate forests—and the resultant peat—capture about one half as much carbon dioxide as do tropical rain forests. Because of year-round heat and rainfall, these trees grow at a far faster rate than those in colder areas. Consequently, replanting the rain forests is very important. Governments of tropical countries such as Brazil, India, and Thailand are making progress in controlling the development of industries using wood products. They, too, are requiring reforestation by logging companies.

Electric utility companies in Holland and other parts of the world are also sponsoring tropical reforestation. The government and power companies of New Zealand are planting 250,000 acres (100,000 hectares) of new trees every year. Major electric power companies in the United States have also begun to sponsor reforestation projects both locally and abroad. Managers of utility companies hope to balance the CO_2 generated by their power plants with the CO_2 absorbed by the trees that they plant. These reforestation projects are also a means to achieve public approval and gain permission to build new and larger power plants.

The amount of all forested land—especially rain forests—is declining every year. The forests are cut for timber, to supply firewood, to increase farmland, and to accommodate expanding urban areas. Although trees are being replanted in many areas, the speed of deforestation is far greater.

Alternative Power Sources

To reduce the generation of carbon dioxide, the use of fossil fuels must be curtailed. Nuclear power, water power, tidal power, and wind and solar energy can be employed to produce electricity that is vital to the economy without producing nitrogen oxides, sulfur dioxide, CO_2, or smoke particles. These alternative sources all have positive as well as negative consequences. However, they all serve as alternatives that curtail atmospheric pollution.

NUCLEAR POWER

For many, nuclear power is an attractive alternative to the burning of fossil fuels. Nuclear power plants produce no sulfur, no nitrogen oxides, and no carbon dioxide.

Nuclear power comes from heat energy generated when the atoms of the metals uranium or plutonium release parts of their substance. This breaking apart is called fission and happens spontaneously. The process is usually carefully controlled in a nuclear power plant. In spite of the accidents at Three Mile Island near Harrisburg, Pennsylvania, in March, 1979, and at Chernobyl village in the Ukrainian Republic, on April 26, 1986, nuclear power is relatively safe. Many people forget that no deaths or cases of radiation disease resulted from the Three Mile Island event. The quantity of radiation released during that incident was smaller than the normal amount of natural radiation.

The Chernobyl accident, however, was a catastrophe. Many plant employees, rescue workers, cleanup specialists, and nearby residents died because of the disaster. Some died quickly, and others after suffering from radiation sickness. Fallout covered thousands of square kilometers in several countries. Farms and villages in the immediate vicinity of the plant were evacuated and remain uninhabited to this day. All agree that the plant was poorly designed, poorly managed, and poorly operated.

The accident began on April 25, 1986. The reactor at Unit 4 was scheduled for a routine shutdown for cleaning and maintenance. Administrators on the day shift planned to conduct tests during the shutdown and turned off some of the safety devices. Then, the shutdown and the tests were delayed because Kiev, the largest city in Ukrainia (now Ukraine), needed more electrical power. A few hours later, the delay ended, and the tasks were resumed by a less-experienced night crew. When the plant was restarted, power generation began to decrease too rapidly. The workers tried to correct the problem and speeded the release of nuclear energy to produce more steam. The disconnected safety equipment did not warn the workers when the reactor began to overheat. Too late, water was fed into the system. The water turned into steam with explosive force. After more confusion, a second explosion lifted the protective cover from the reactor and allowed air to enter. An intense fire resulted, and clouds of radioactive materials were sent into the air. Chaos followed.

World opinion about the use of nuclear energy has been strongly affected by these two crises. The incident at Three Mile Island made it impossible to obtain public support for additional nuclear power plant construction in the United States. The accident at Chernobyl convinced many people in many countries that the advantages of nuclear power were seriously overshadowed by the risks.

However, several countries, including France and Japan, are heavily dependent on nuclear power for the bulk of their electricity and continue to build new plants. The leaders of these countries are neither ignorant nor reckless. They have carefully studied all facets of the problem. Indeed, the Japanese are highly

This interior view of a tokamak shows the size of the donut-shaped cavity, or torus, in which the hot nuclear plasma can enter into a nuclear fusion reaction. (Courtesy of the Princeton Plasma Physics Laboratory)

sensitive to the dangers of atomic energy because of their experience with the atom bomb.

Today, high-tech nuclear power plants are designed to eliminate most of the errors attributed to machine malfunction or human mistakes. The new plants produce few pollutants. Although most environmental problems have been solved, one enormous difficulty remains. This challenge involves uranium rods—the fuel used to operate the power plant. When uranium rods are placed in the reactor core, they support a chain reaction that produces heat to make steam. The steam drives the turbines to generate electricity, which is distributed to commercial and domestic users in the area. The manufacturing process goes smoothly except for one detail. No one can decide where to dispose of the used atomic fuel rods.

When uranium rods have served their purpose and can no longer be used as fuel, they must be replaced. At present, the rods are removed from the reactor core and stored underwater in trenches at the site of the power plant. However, the used rods remain radioactive. For 40 years, officials of the U.S. Department of Energy have spent vast amounts of money searching for a safe, acceptable location to bury those spent—but highly dangerous—fuel rods. The most suitable location seems to be in the mountains of central Nevada. Naturally, the people of Nevada are not pleased to be the hosts for nuclear materials. Long legal battles are likely if the federal government chooses central Nevada as the final resting place for radioactive waste. Meanwhile, the amount of spent fuel continues to increase.

In contrast to nuclear fission, producing electricity by the process of nuclear fusion has greater appeal. Fusion is the same nuclear process that takes place in the sun and that produces the heat that sustains life on Earth. The merging (or fusion) of two

This exterior view of a tokamak shows the complicated wiring needed to generate the magnetic field that holds the hot nuclear plasma in place. (Courtesy of the Princeton Plasma Physics Laboratory)

hydrogen atoms to form one helium atom generates a great amount of heat energy. The heat can be used to produce steam. Hydrogen—the only ingredient needed to begin the process—is in limitless supply. Helium—the result of the process—is harmless and nonpolluting. However, a system to fuse the hydrogen atoms requires costly apparatus and huge amounts of electricity. So far, the electricity required to run the system exceeds the quantity produced. Consequently, there is a net loss. Much more research is needed to achieve a practical system for generating electricity from nuclear fusion.

WATER POWER

For untold ages, the power of water was used to perform simple tasks. A wheel placed in a swiftly moving stream or river was turned by the force of the rushing water. Waterwheels often provided the energy to grind grain into flour. Today, most water power is generated by large dams built to control the flow of a river. A dam is constructed across a narrow river valley, blocking the river's course. A lake is formed behind the dam. The surface of the lake is higher than that of the natural riverbed. When the water is released, the torrent flows downstream through large pipes connected to giant turbines within the dam. The rushing water spins the turbines that rotate the electric generators. The water-generated electricity—called hydroelectric power—is produced without the use of fossil fuels.

Unfortunately, this environmentally safe procedure has some negative side effects. When the flow of a river is halted by a dam, silt, sand, and clay normally carried by the water settle out and accumulate in the lake. After a while, the silt must be dredged (removed) from the lake. Dams also interfere with the movement of fish, especially salmon. In the Northwest, salmon breeding is impaired, and the salmon population has declined in recent years. Scientists and environmentalists are seeking solutions to this problem.

Hydroelectric power accounts for about 8% of the electricity generated in the United States. Although additional expansion

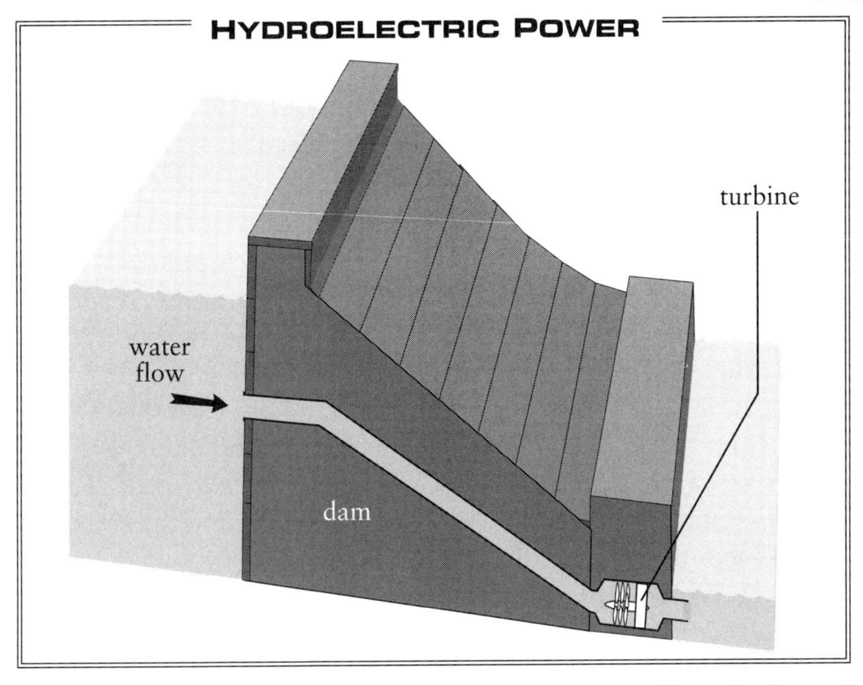

The sketch shows an internal cross section of the dam and how the flow of water is directed to a turbine that drives an electric generator.

is possible, most of the better dam sites are already in use. Flat or gently rolling land is not suitable as the site of a dam. In order to generate the rush of water necessary to rotate the turbines, the level of the lake must be much higher than the level of the riverbed. Also, level land along a river is frequently very fertile. The flooding caused by dam construction can take important agricultural acreage out of production.

The Electric Consumers Protection Act (1986), states that water-powered electric plants must undergo an environmental impact analysis before gaining a license renewal. Preservation of wildlife habitats and consideration of agricultural irrigation have limited the upgrading of some existing plants. In the United States, electricity produced by hydroelectric plants has actually declined since 1986.

On a worldwide basis, water power has expanded rapidly since the 1950s. However, other means of generating electricity have expanded even more rapidly. Therefore, the proportion of electricity produced by water power has declined to about 20% of worldwide production.

Opportunities for large-scale expansion of hydroelectric power facilities are present in many developing countries. However, the issue of environmental protection has caused leaders of these countries to be cautious about new projects. Today, officials from industrialized countries are considering new programs to help develop and finance water-power facilities in those areas.

To support such developments, scientists and engineers have begun research programs such as surveying prospective dam sites. Fortunately, recent technology has facilitated the construction of economical, small-scale, hydroelectric stations in remote areas. A World Bank survey covering 31 developing countries reveals that power generation in these settings has doubled over the past 10 years.

TIDAL AND OCEAN WAVE POWER

The power of ocean tides is considerable. The rush of incoming and outgoing tides generates the same amount of power as the work of countless dams. If this enormous force could be harnessed in an economical manner, tidal power would be another clean source of energy. For the past 20 years or so, tidal power has received enthusiastic publicity. At present, however, there are only two major tidal power facilities in the world. One is located on the northwestern coast of France and the other in Canada on the Atlantic coast of Nova Scotia.

A special kind of dam called a "barrage" is used to capture the power of the tides. As high tide comes in, gates in the barrage are opened. The rush of water rotates the turbines at the sides of the gates. When the tide is fully in, the gates are closed, and the water is trapped until low tide. At low tide, the water within the barrage is higher than the water beyond. The gates are reopened, and water rushes back toward the ocean. The torrent of water again turns the wheels of the turbines. Therefore, the turbines generate electricity during the surge of both high and low tides.

The production of electricity by tidal power is limited to 8 or 10 hours each day. The facilities are expensive to construct and operate. In order to justify such an investment, the tides at the site must rise at least 15 feet (5 meters) or so. Since most tides range between 3 and 4 feet (about 1 meter), most seashore areas are unsuitable for such projects. Technical experts predict that no more than 1% of the world's power needs can be provided by the tides.

Many different devices have been invented to take advantage of the untapped energy generated by ocean waves. So far, these methods have produced electricity on a relatively small scale. However, the concept appeals to inventors who see wave power as free and nonpolluting. Consequently, research programs of varying size and scope are under way in Japan, Norway, Denmark, and India as well as in the United Kingdom and the United States.

The difference between the temperatures of surface water and deep water has been considered a possible source of energy. Although the idea was conceived in the 1880s, no project was undertaken until 1930. Unfortunately, that system consumed more electric power than it generated. A second project was planned for a ship anchored off the coast of Brazil. The work was abandoned when the ship was wrecked by a storm. At present, the U.S. Department of Energy and other organizations are doing research on the subject.

Although ocean power is freely available, the techniques to capture this power are very complex. Indeed, the power of the oceans will not be harnessed until conventional sources become far more expensive.

GEOTHERMAL POWER

A layer of intensely hot, partially melted rock called magma lies below the crust of the earth. In some locations, narrow faults in the crust allow water to seep down near the magma, heat to the boiling point, and rise to the surface of the earth as steam. This

Geothermal energy is captured by this complicated machinery. The heat is used to make steam that spins a turbine that drives an electric generator. (Courtesy of Warren Getz and the National Renewable Energy Laboratory)

is the origin of the hot springs and geysers found in Yellowstone Park. In places such as Iceland the faults are common, and many hot springs have been formed. Their steam is harnessed to drive electric generators. Some homes in Iceland are heated directly by water piped from the hot springs.

Large geothermal electric plants are expensive to build and operate. In addition, few areas are suitable. Geothermal energy will never account for more than a modest proportion of all energy production.

WIND POWER

Windmills generate energy to perform several kinds of work. Modern windmills are fairly high structures with blades attached to a shaft near the top. The blades are driven by the wind and power various kinds of machinery.

Archaeologists believe that the Babylonians built windmills as early as 1700 B.C. However, the first documented windmills were constructed in Persia (Iran) in 644 A.D. They were strange structures with the blades set parallel to the ground and attached to a short shaft. The shaft was connected directly to a small millstone used to grind grain into flour. Standard or vertical windmills were designed to copy the action of water mills. The wind-driven blades are attached to a horizontal shaft that is connected by gears to another shaft. This shaft activates a variety of devices. About 900 years ago, windmills became a popular source of power in Europe. The famous windmills of Holland are used to mill grain and to pump water from low-lying areas. Windmills have been used to drive machines such as the looms that weave fabric. Since 1890, they have been used to generate electricity. This application was first tried on the flat plains of Denmark.

In the United States, lumber companies used windmills to run sawmills. Farmers in the Midwest have used windmills to generate electricity and pump water from deep wells since the 1920s. Most windmills were relatively ugly structures of galvanized iron. Each stood about 25 feet (7.5 meters) high and was topped with a multibladed fan. They creaked and groaned in the high prairie winds as they pumped well water for domestic use and into troughs for farm animals.

The petroleum shortage of the 1970s made everyone conscious of the dependence on Middle Eastern oil. This problem caused some inventors to think about harnessing wind energy. The systems developed during the oil crisis are far more elegant than the old farm windmills.

Top officials of the U.S. Department of Energy are encouraging the use of windmills to generate electric power in two different settings. On privately owned rural or semirural property, officials hope that owners of land not connected to power lines will attempt to generate their own electricity. Then, utility companies will not be required to extend expensive transmission equipment.

Even if the rural site is connected to power lines, a windmill system can conserve or replace fossil fuels. The property owner can generate part of the needed electricity by using a windmill

This type of windmill has long been a common sight on midwestern farms, where it is used to drive a pump that brings up water from an under ground well. (Courtesy of the U.S. Department of Agriculture and the National Renewable Energy Laboratory)

and buy the rest from a utility company. In some places, surplus wind-generated electricity can be sold to a power company.

Indeed, many different arrangements are possible. The windmill owner can achieve complete energy independence by storing surplus energy in batteries. Heavy-duty battery packs, similar to those used to power golf carts, hold enough electricity to last three or four days. Diesel engines or ordinary gasoline engines can be used as emergency electric generators.

Conditions must be just right for the household windmill to be a wise investment. If electricity rates are high, purchasing and maintaining the equipment may be economically sound. However, the site must be very windy, and the system must be tall enough (about 30 feet or 9 meters) to catch the wind. In addition, the location must be far enough from neighbors so that the noise will not be a nuisance. If these conditions are met, each privately owned windmill will lower the output of carbon dioxide by two tons per year. That is the amount of CO_2 generated by a conventional power company to produce a family's yearly electrical supply.

The most important use for windmills is a wind-driven power plant that takes the place of a conventionally fueled utility company. A large number of windmills are erected on pylons that stand 40 feet (12 meters) or higher and are spaced about 100 feet (30 meters) apart. Each windmill can generate as much as 500 kilowatts of power. When many windmills are linked together into a power grid, they can supply electricity to a modest-sized community and eliminate the need for a fossil-fueled power plant.

A site for a windmill farm should be at least 50 acres (20 hectares) in size. However, the land under the pylons need not be idle. It can be cultivated for crops or used as pasture land. Although the windmills produce some noise, the sound is not much greater than that of a busy office. One drawback is the danger to birds. They do not always see the whirling blades and fly into them. Scientists from the U.S. Department of Energy and the National Audubon Society are working on this problem.

An array of modern windmills with giant blades. These windmills can generate electricity for whole communities. (Courtesy of the National Renewable Energy Laboratory)

SOLAR POWER

Scientists have devised two types of solar energy systems. The older system uses the sun to directly generate heat. Such systems can be quite modest. A series of curved reflectors focus the sun's rays on a central pipe that holds water or some other fluid. Heated water can be used directly for household purposes such as bathing and washing dishes.

Slightly more elaborate solar energy systems are used to store heat. The heated fluid is passed through many channels in a large ceramic block. Such blocks will hold heat for several hours, so there can be hot water after the sun sets. By eliminating the need for a hot water system that uses electricity or natural gas, a small reduction in carbon dioxide generation is achieved. Some larger installations have been adopted by home owners to provide home heating in the winter. However, the need for home heat is greatest when the sunshine is weakest and the cloud cover is most extensive.

Alternatives to providing heat to individual homes have been adopted by some electric utility companies on an experimental basis. These companies use large numbers of mirrors focused on a target container. The target material, which is often metallic sodium, is made very hot by these concentrated rays. When that material is passed through a device like a radiator, it can heat water to make steam. Then, the steam is used to drive a turbine that turns an electric generator.

These concave mirrors focus the sun's rays on the tube of liquid in the center location. The hot liquid can transfer its heat to water in order to make steam. (Courtesy of Warren Gretz and the National Renewable Energy Laboratory)

The other type of panel is used to generate electricity directly from the sun's rays. This process, called the photoelectric effect, uses panels that are about 3 feet (1 meter) wide and 6 feet (2 meters) high. Each panel faces the sun and is divided into hundreds of small cells. Each cell contains a thin layer of silicon on top of a thin layer of metal. These layers are backed by a ceramic wafer called a semiconductor. When sunlight strikes the silicon, it jars some electrons from the silicon atoms. The electrons move through the metal layer into and through the semiconductor, and they then flow into an electric wire behind the semiconductor. This electron flow produces an electric current.

This is the technology used by satellites, space stations, and the robot landers that go to other planets. It is very clean and safe and can continue to work without moving parts or any

This large array of mirrors focuses the sun's energy on the container atop the central tower. The heat is used to make steam. (Courtesy of Sandia National Laboratories and the National Renewable Energy Laboratory)

These panels contain tiny silicon wafers that release electrons when energized by the sun's rays. When linked together, an electric current is generated with no moving parts. Such panels are used to power satellites, space stations, and robot interplanetary explorers. (Courtesy of the 3M Corporation and the National Renewable Energy Laboratory)

adjustments for years. However, each panel generates only a small amount of electricity.

At present, commercial solar-generated power has been limited to medium-sized plants located in sites in the deserts of the southwestern United States and a few sunny countries. The electricity produced is relatively expensive because of the high cost of the panels. However, work goes forward on developing better methods to tap the sun's energy.

Hydrogen as Fuel

Some clean sources of energy such as tidal power, wind power, and sun power produce electricity on a variable schedule. The tides run twice a day. The winds are undependable. The sun cannot generate energy on cloudy days or at night. Hydrogen, another clean energy source, is always available. Hydrogen can be burned to serve any purpose requiring heat—including the production of electric power.

Hydrogen (H) can be made from ordinary water (H_2O) by the use of electricity. When an electric current is passed through the water, the hydrogen (H) portion of the molecule is separated from the oxygen (O). Since both oxygen and hydrogen are gases at room temperatures, they bubble up and break through the surface of the water. The gas can be captured in containers and stored for later use.

General acceptance of hydrogen-generated power will take many years. Industrialized countries have a huge investment in the production of energy by fossil fuels. Companies that mine and transport coal, explore for oil and natural gas, dig wells, and build pipelines would be financially weakened by a switch to hydrogen. Many other industries would be adversely affected. All the furnaces and boilers used to make steam for turbine generators would become obsolete. Automobile, truck, and tractor engines would be worth less than they are worth now. Indeed, huge investments would be lost if fossil fuels were replaced by hydrogen.

At present, hydrogen is about four to five times as expensive as methane or natural gas. However, it generates about three times as much energy per pound. Engineers and economists expect that hydrogen-produced energy will be priced competitively in about 50 years.

The use of such an alternative power source will be especially advantageous to many developing countries. Many of these areas are near the equator, where solar energy can be harnessed more consistently than in temperate countries. In addition, those nations have less capital invested in industries devoted to the use of fossil fuels. Consequently, fewer economic barriers would hinder the conversion to hydrogen-generated power. If the leaders of developing countries focus on developing hydrogen technologies, their countries could lead the movement to achieve bountiful, pollution-free energy.

There are many ways to reduce or control atmospheric pollution. Some require a reduction in industrial productivity and, perhaps, a decrease in general prosperity. Others might involve major changes in lifestyle.

The ways that would cause the least disturbance to the economy or to individual lifestyles are the alternative energy sources that could substitute for fossil fuels. At present, most alternative energy sources cost more than fossil fuels. However, surveys reveal that people will accept higher prices for utilities and manufactured products if the net outcome is a cleaner, less polluted atmosphere.

Glossary

acid rain Rainfall having more than normal acidity, most often generated by oxides of sulfur and nitrogen in the atmosphere.

aerosol An assembly of very small droplets of liquid that are light enough to float in the air.

air mass A large body of air that has relatively uniform features. Within the mass, temperature and barometric pressure are similar.

albedo The amount of reflected energy compared to the total energy received by a surface; usually expressed as a percentage.

anticyclone A large air mass characterized by high barometric pressure with winds moving in a clockwise direction in the Northern Hemisphere and counterclockwise in the Southern Hemisphere.

barometric pressure The weight of a column of air above a specified area; measured in centimeters of the height of a column of mercury forced upward in a glass tube from which the air has been removed.

coalescence The merger of two small water droplets to form a larger droplet.

cold front The boundary between a mass of cold air and a mass of warm air whereby the cold air usually underruns the warm air, pushing it up.

condensation The process by which a vapor becomes a liquid or a solid.

convection Movement of material away from a heat source toward a cooler region. Heat energy is transferred by such movement.

convergence The narrowing of a flow, as when water runs down a drain.

cyclone An area of low barometric pressure in which the winds blow in a counterclockwise direction in the Northern Hemisphere and clockwise in the Northern Hemisphere.

depression Another name for a cyclone.

dew point The temperature at which water commences to condense from the air onto a solid surface; the point at which the relative humidity reaches 100%.

eddy Small shift in the movement of air currents.

evaporation The process by which a liquid is changed into a gas.

fossil fuels Fuels that are the remains of once living matter: coal, oil, and natural gas.

front The border or boundary between two air masses.

greenhouse effect The action of gases in the atmosphere that retain and re-emit radiant heat that would otherwise escape into space.

humidity A measure of the amount of water vapor in the atmosphere.

infrared Light that carries heat energy and is not visible to humans.

inversion A layer of warmer air on top of a layer of cooler air.

isobar A line connecting two points on a weather map that have the same barometric pressure.

jet stream A strong flow of air in the stratosphere that moves from west to east in the Northern Hemisphere and that often marks the boundary between polar air and tropical air.

latitude The distance north or south from the equator measured in degrees of the circumference of a circle.

meteorology The study of the earth's atmosphere with particular emphasis on weather forecasting.

monsoon A seasonal wind usually accompanied by heavy rains.

relative humidity The amount of water in the atmosphere relative to the maximum amount that could be held at the prevailing temperature—expressed as a percentage.

semiconductor a device that allows an electric current to flow in one direction only.

smog Originally, a mixture of smoke and fog; now also refers to a mixture of ozone, oxides of nitrogen, and particles made up of carbon compounds.

squall line The boundary between converging air currents where heavy, sudden rain storms erupt.

stratosphere The layer of atmosphere above the troposphere where the air is very thin and calm conditions are normal.

synoptic The summarization of much weather information, often in graphic format as a weather map.

thermal a relatively restricted and well-bounded column of rising air over a warm area.

troposphere The lowest layer of the atmosphere, from the surface to a height of from 6 to 12 miles (10 to 20 kilometers).

warm front The boundary between an advancing mass of warm air sloping upward from the ground and pushing into a mass of cold air.

wind The movement of air relative to the surface of the earth, described in terms of the location of origin so that a northeasterly wind comes out of the northeast.

wind shear Sharp change in the direction or the speed of the wind.

Further Reading

Allaby, Michael. *Dangerous Weather.* New York: Facts On File, 1997–98. A six-volume set on weather. Titles include: *Blizzards, Droughts, Floods, Hurricanes, Tornadoes,* and *A Chronology of Weather.*

Baines, John. *Conserving the Atmosphere.* Austin, Tex.: Steck-Vaughn, 1990. This book focuses on the causes of various forms of air pollution. Carefully written and well illustrated, it contains a useful listing of foundations and other organizations dedicated to the preservation of the environment.

Blashfield, Jean F., and Wallace B. Black. *Global Warming.* Chicago: Children's Press, 1991. This title includes sections on hands-on experiences and experiments that can be conducted in a classroom setting.

Lambert, David, and Ralph Hardy. *Weather and Its Works.* New York: Facts On File Publications, 1984. Written and first published in Great Britain, this volume has a truly global orientation. The authors do a particularly good job of distinguishing among the various climate types around the world. This book is now out of print; however, it may still be in libraries.

Mason, John. *Weather and Climate.* Englewood Cliffs, N.J.: Silver Burdett Press, 1991. This book is written simply, with plentiful illustrations. Its broad coverage includes most

aspects of atmospheric pollution and the technical means for observing changes in the atmosphere.
Miller, Christina G., and Louise A. Berry. *An Alert: Rescuing the Earth's Atmosphere.* New York: Simon & Schuster, 1996. A young adult book on air pollution and political solutions.
Schaefer, Vincent J., and John A. Day. *A Field Guide to the Atmosphere.* Boston: Houghton Mifflin Co., 1983. One of the well-known series of Peterson Field Guides, this copiously illustrated handbook concentrates on helping the reader distinguish between different types of clouds and between different types of snow and other forms of precipitation. The first author was a pioneer in weather modification.
Weather and Climate. Alexandria, Va.: Time Life Books, 1991. This heavily illustrated book mainly presents explanations of the how and why of weather and climate effects.

Index

Page numbers in *italics* indicate illustrations.

D

E

F

G

H

I

P

R

S